The European Environmental Conscience in the EU

This book analyses how sustainability affects internal decision-making within the European Union and its external relations in working towards achieving its long-term goal of a climate-neutral Europe by 2050.

Applying the term "European environmental conscience" as the perception of environmental degradation leading to a growing public awareness of the issues, a notion of common responsibility, and European institutions dealing with these growing concerns, the book investigates its emergence as a lever for deeper European integration and in fostering a genuine European identity. Examining policy areas such as green finance, innovation policies, and foreign policy, it reveals the impact these concerns have for other policy fields.

This book is of key interest to scholars and students of environmental economics and politics, sustainability governance, green finance, climate policy, energy policy, and more broadly, to European studies and international relations.

Marjorie Tendero is an Associate Professor of Economics at ESSCA School of Management, Angers, France, and a member of the EU*Asia Institute. She is an associate researcher at the SMART(Structures and Markets in Agriculture, Resources and Territories) laboratory.

Christoph Weber is an Associate Professor of Economics at ESSCA, School of Management, Lyon, France. He is the Deputy Director of the EU*Asia Institute at ESSCA School of Management.

Routledge Studies on the Governance of Sustainability in Europe

Series editors:

Thomas Hoerber
ESSCA School of Management, France

Jenny Fairbrass
University of East Anglia, UK

Kristina Kurze
University of Göttingen, Germany

Helene Dyrhauge
Roskilde University, Denmark

At the beginning of the twenty-first century the EU is faced with major economic, environmental and societal challenges. Climate change is at the forefront of these challenges having a wide-ranging impact on the economy and society. This series will examine tensions and ambiguities in European sustainable development from an interdisciplinary and multidisciplinary background, covering the impact of sustainable development policies on European integration, and sustainability issues in energy, environmental, transport, development and agricultural policies.

Making the European Green Deal Work
EU Sustainability Policies at Home and Abroad
Edited by Helene Dyrhauge and Kristina Kurze

The EU in a Globalized World
Edited by Thomas Hoerber, Alexandre Bohas, and Stefano Valdemarin

EU Energy and Climate Policy after COVID-19 and the Invasion of Ukraine
Decarbonisation and Security in Transition
Edited by Matúš Mišík and Andrea Figulová

Economic Aspects of the Sustainability Transition in Europe
Rosa Maria Fernandez Martin

The European Environmental Conscience in the EU
Finance, Innovation, and External Relations of the EU
Edited by Marjorie Tendero and Christoph Weber

For more information about this series, please visit: https://www.routledge.com/Routledge-Studies-on-the-Governance-of-Sustainability-in-Europe/book-series/GSIE

The European Environmental Conscience in the EU

Finance, Innovation, and External Relations of the EU

Edited by Marjorie Tendero and Christoph Weber

Routledge
Taylor & Francis Group

LONDON AND NEW YORK

First published 2025
by Routledge
4 Park Square, Milton Park, Abingdon, Oxon OX14 4RN

and by Routledge
605 Third Avenue, New York, NY 10158

Routledge is an imprint of the Taylor & Francis Group, an informa business

© 2025 selection and editorial matter, Marjorie Tendero and Christoph Weber; individual chapters, the contributors

The right of Marjorie Tendero and Christoph Weber to be identified as the authors of the editorial material, and of the authors for their individual chapters, has been asserted in accordance with sections 77 and 78 of the Copyright, Designs and Patents Act 1988.

British Library Cataloguing-in-Publication Data
A catalogue record for this book is available from the British Library

ISBN: 978-1-032-63668-9 (hbk)
ISBN: 978-1-032-65636-6 (pbk)
ISBN: 978-1-032-65635-9 (ebk)

DOI: 10.4324/9781032656359

Typeset in Times New Roman
by SPi Technologies India Pvt Ltd (Straive)

Contents

Figures

Tables

Contributors

Paiman Ahmad is an Assistant Professor, University of Raparin, Rania-Kurdistan Region-Iraq.

Alhamzah Alnoor is a senior lecturer at the Southern Technical University in Basra, Iraq.

Tomasz Braun is an academic and a practicing lawyer. He is Pro-Vice-Rector for International Relations at Lazarski University, Director of the Institute of American Economy and Transatlantic Relations, and university lecturer in Poland and France and a guest lecturer in Italy, the UK, Finland, Austria, Germany and the Netherlands.

Gaël Callonnec has a PhD in economics and is in charge of coordinating work on the macroeconomic impact of environmental tax, energy, and energy transition policies at ADEME (French Environment and Energy Management Agency).

Naciba Chassagnon-Haned is a Professor of Economics, Research Director, and Associate Faculty Dean of ESSCA School of Management, Lyon, France.

Daniel Feser is a Professor of Risk and Sustainability Sciences in the Department of Social Sciences and the Society for Institutional Analysis (sofia) at Darmstadt University of Applied Sciences, Germany.

Mathieu Garnero is Head of Economics and Finance at ADEME (Agence de l'Environnement et de la Maitrise de l'Energie).

Dejan Glavas is an Associate Professor of Finance and Director of the AI for Sustainability Institute at ESSCA School of Management, Paris, France.

Amina Hamani is a Research Associate at the Nimec Laboratory at the University of Caen and a lecturer at the ESSCA School of Management.

Noam Léandri is an Economist-statistician of INSEE, former Secretary General of ADEME (French Environment and Energy Management Agency), research associate at the EU*Asia Institute at ESSCA School of Management.

Florent Junior Okala Onana is a Financial Security Analyst at BNP Paribas.

Chiara Pappalardo holds a Doctor of Juridical Science (S.J.D) degree from Georgetown University Law Center. She is licensed to practice law in Italy and is a former visiting research scholar and professional lecturer in law of the Washington College of Law at American University in Washington D.C.

Frauke Pipart is Head of Research at the Centre for Multilateral Negotiations (CEMUNE). She holds a PhD in Political Science from the University of Louvain.

Marjorie Tendero is an Associate Professor in Environmental Economics at ESSCA School of Management and Associate Researcher at the SMART laboratory (Structures and Markets in Agriculture, Resources and Territories).

Maxence Tétard is a Decarbonisation Project Manager at BRAINERGIES.

Muhammad Usman is associated with the School of Economics and Management and the Centre for Industrial Economics, Wuhan University, Wuhan, China.

Christoph Weber is an Associate Professor of Economics at ESSCA, School of Management, Lyon, France. He is the Deputy Director of the EU*Asia Institute at ESSCA School of Management.

Acknowledgements

First of all, I would like to thank my co-editor, Marjorie Tendero, for the fantastic collaboration on this book project and for her outstanding contributions. I would also like to thank the contributors to this edited book, without whom this book would not have been possible.

Second, I would like to thank Thomas Hoerber for guiding us through the book project. I would also like to thank Silke Leukefeld for her help in organising the conferences on which the book project was based. And David Rees for proofreading.

Third, special thanks to the series editor and the team at Routledge for helping us at every stage of the book project.

Fourth, I would like to thank my school, the ESSCA School of Management, for supporting the book project.

Finally, I would like to thank my family for their support and encouragement throughout this process.

Christoph Weber
Lyon, July 2024

It's always a pleasure to write these lines. A heartfelt thank you to Thomas Hoerber, without whom this collaboration with Christoph Weber would never have happened. In this journey, you have been the "Grand Master Yoda" to my developing path as a researcher. I would also like to thank my co-editor, Christoph Weber. His insights and dedication were essential in bringing this project to life. It has been an honour to work with you.

I would also like to extend my gratitude to all the contributors of this book, whose expertise made this book possible. Special thanks to Silke Leukefeld for her assistance in organising conferences that facilitated valuable exchanges among contributors and for her efforts in sustaining the vibrant community at the EU-Asia Institute. Your organisational skills and enthusiasm are invaluable.

I am also grateful to those involved in the editorial process at Routledge, who supported us at every stage. A special thanks to the anonymous reviewers of our proposal.

I thank my colleagues at the EU-Asia Institute for their encouragement. My gratitude extends to ESSCA School of Management for its support.

As we close this book, I'm reminded that in the field of environmental conscience in Europe, "there is not try, only do".

Marjorie Tendero
Angers, July 2024

Introduction

The European Environmental Conscience in EU Economics

Marjorie Tendero and Christoph Weber

The European Union (EU) has its roots, in large part, in securing energy provision: it started in 1951 with the European Coal and Steel Community (ECSC). Steel production was an essential part of the economic output which required a substantial amount of energy. As France needed more coal than it produced itself, the idea of bundling coal production in a supranational organisation like the European Steel and Coal Community was very appealing to France as it helped secure coal provision on the one hand, and also domesticated Germany by having an influence on their industrial production on the other (Hörber, 2006).

However, at that point in time, the focus was not on the environment, reducing energy consumption, or sustainability. The vision was to reindustrialise Europe and maintain the necessary level of energy provision to accommodate economic recovery and growth (Sarti & St. John, 2019). The idea that economic growth without any side effects is not the norm but that it has to be accompanied by environmental policies dates from the late 1960s and 1970s with the oil crises. This is the time that Hörber (2012) considers as the beginning of the formation of a European environmental conscience (Hörber, 2012). Growing concerns about forest dieback, acid rain, nuclear power, oil prices, and other problems led to a more thorough contemplation of parts of society about the way of living and the way to go forward. In most Western European countries, the 1970s mark the beginning of some visible protests of environmental movement organisations (Rootes, 2003). This led eventually to some serious discussions about the topics of environment and sustainability. Two major events were the implementation of the first "Programme of Action of the European Communities on the Environment" declared in November 1973, and the first World Climate Conference in 1979. Until now, 26 other climate conferences have followed, and the EU has set up its eighth Environmental Action Programme.

Even so, the process of a full commitment of the EU towards sustainability or sustainable development has been rather long. In 1986, the Single European Act marked the first time that the EU included "environmental protection integration" which demanded a cross-check of all sectoral policies with respect to

DOI: 10.4324/9781032656359-1

environmental goals. In 1992, the Maastricht Treaty brought some pledges to the goal of sustainable development – not just economic development. Then, in 1997, the Amsterdam Treaty introduced a specific article which asked for a clear consideration of environmental principles for all actions and policies of the EU (Barnes & Hörber, 2013).

At the beginning of this development, the focus was mainly on energy. In fact, Hörber and Weber (2022) identify the relationship between energy and environment as a lever for deeper European integration. This can be explained by the fact that many environmental problems are not restricted to a specific region or territory but have cross-border effects.[1] Hörber and Weber (2022) go even further by concluding that European integration in the field of environmental politics and sustainability fosters a genuine European identity. Notably, the EU now appears as a major actor in multilateral governance as shown by multiple cases of involvement of the EU in the United Nations Environmental Governance (Wouters et al., 2012). Furthermore, the EU is involved in the establishment of numerous multilateral agreements regarding environmental policies (Delreux, 2018; Vogler, 2005). Another important platform for making progress in environmental regulation is the Committee on Environmental Policy of the United Nations Economic Commission for Europe (UNECE). Several propositions of this committee have led the way for EU regulation (Razzaque, 2012). At the same time, the actions of the EU have also motivated other non-EU countries to implement environmental policies and to ratify treaties of the UNECE (McNeill, 2021; Schulze & Tosun, 2013; Shyrokykh, 2022). Leading by example – in spite of the fact that the EU is considered to be a "leaderless leader" (Oberthür & Roche Kelly, 2008) – and fostering local climate initiatives have promoted the EU as a global leader in multilateral climate governance (Jänicke & Wurzel, 2019).

The remainder of this introduction is organised as follows. The second section will detail the importance of finance for economic transition and present some major achievements in terms of financial regulation in the EU. The third section will expand on the importance of innovation for economic transition towards sustainability and will elaborate on the main tools of the EU in the field of innovation policy. The fourth section will demonstrate the significance of external relations in achieving a sustainable transition at the EU level. This section will also provide a summary of European external relations policies and outline their links to finance and innovations. This section will also explain how external relations may impact sustainability. This is followed by a discussion of the theoretical framework for this edited book: the notion of the European environmental conscience. The underlying theoretical framework serves as glue by sticking the individual chapters together.

The conclusions at the end of the book will summarise the main findings of the individual chapters and answer the question of whether the European environmental conscience has continued to drive European integration. Assuming that the authors of the selected chapters do indeed find evidence that there is a growing European environmental conscience, this can guide the way for further development in the EU. This is even more pressing in a situation where Europe is under outside pressure from Russia – both in a military and an economic sense. The necessary transformations – for instance, in the energy sector – due to the wish to increasingly decouple the EU from Russian influence, open a great opportunity for change that can serve several goals (integration, autonomy, and sustainability). To make this happen, connecting and identity-building a new mindset among Europeans would be a major driving force for a successful transformation of Europe.

Finance

The Important Role of Finance in Economic Transition

Regulation plays a dominant role in pushing the economy and society towards sustainability. Apart from regulatory policy, the state also affects the economic transition of countries in numerous ways. If we assume that innovation and transition are necessary, then we can make the case for the relevance of the state. An example might be the provision of adequate infrastructure (charging stations for vehicles, railway lines, bicycle lanes), other public goods (mainly education or financing basic science), or subsidies (subsidies for power-to-liquid production[2] or energetic renovation of real estate) which facilitate the transformation of the economy towards more sustainability.

However, huge investments are necessary to transform the economy towards sustainability. Therefore, most investments and research and development (R&D) expenditure will have to come from the private sector. Figure 0.1 gives a quick snapshot of the situation in OECD countries. The figure shows the investment shares of the public, private, and corporate sectors. The yellow dots indicate the total investment of the country as a percentage of GDP, and the green triangles show government spending as a percentage of GDP. All values are average values of the period 2000 to 2010. What we can see in this graph is that among all these countries, government investment accounted, on average, for roughly 16% of all investments. Among EU countries, the values range from 10.9% for Germany to 25% for Greece. In fact, the portions have even decreased in recent years – from 18.4% in 2010 to 15.6% in 2019. The numbers for government investment that are reported here overstate reality since government investment includes part of spending for military purposes. Thus, the bulk of investment comes from the non-governmental sector. In most countries, the corporate sector plays the most important role in investment activities (on average, 60% of total investment). In the EU, Greece has the lowest contribution of the corporate sector (37.9%) whereas Sweden's corporate sector is responsible for 68.6% of all investment. The household sector contributes, on average, 24% to total investment with Sweden at the lower end (12.6%) and Greece at the higher end (36.2%). However, private investment has also stagnated, and savings in the Eurozone have not been channelled towards effective investment (Cozzi, 2016).

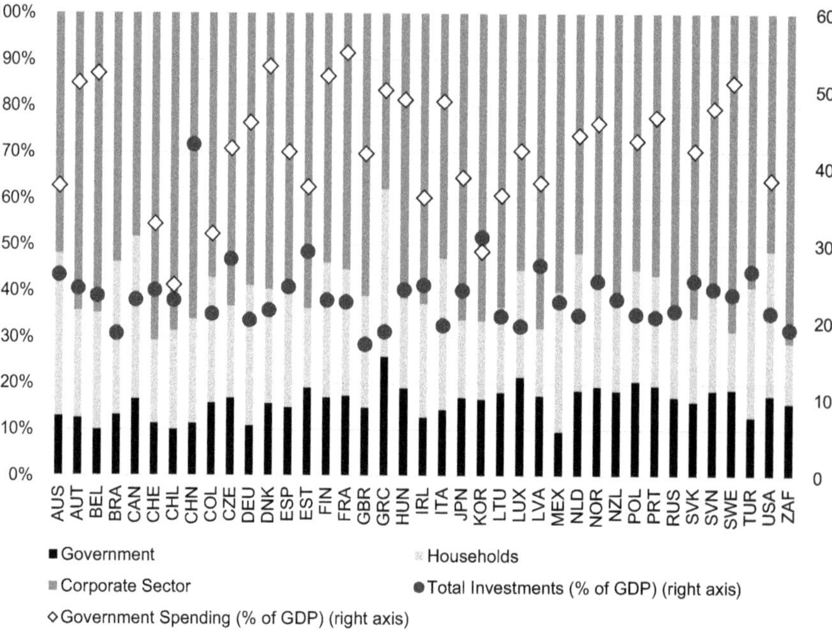

Figure 0.1 Contributions of different sectors to total investments (2000–2020).

This leads to the question of whether a larger government sector (as measured by government expenditure as a percentage of GDP) is beneficial for investment. The quick answer is no, since there is no positive correlation between government spending and investment. For instance, France has the largest government sector among EU countries, with spending of 54.9% of GDP, and the country has investments worth 22.6% of GDP. Estonia reached an investment rate of 29.0% with government spending of only 37.4%. These are not outliers, but cases that fit the global picture. To complete the story, we can state that a higher share of corporate investment is positively related to total investment as a percentage of GDP. To summarise, the investment that is necessary for economic transformation towards sustainability will mainly have to come from the private sector. This makes the topic of finance important for reaching environmental goals. However, the regulation of finance plays a very important role in driving investment towards more sustainability. This will be discussed in several chapters of the book. We will now discuss how financial regulation has evolved in the EU and when sustainable finance became a topic.

A Quick History of Financial Regulation in the EU

The first ideas of a joint economic market were already set out in the Marshall Plan, and were also proposed by Jean Monnet (Kindleberger, 2015). The then Dutch foreign minister, Jacques Beyen, proposed the idea of a common market in 1952, and it was formally proposed during the Messina conference in 1955, followed by the Spaak report in 1956 (Dedman, 2010). The Treaty of Rome (ToR) of 1957 then introduced the freedom of capital as one of the four freedoms of the European Economic Community (EEC). Article 26 of the ToR mentioned that the free movement of capital was one of the main components of the internal market, and Articles 67–73 of the ToR detailed the regulations of the free movement of capital (Craig & De Búrca, 2020, 2021). This led to a sizeable increase in US foreign direct investment in the EEC and to mergers among corporations in the industry sector (Dedman, 2010). The ToR also stipulated the liberalisation of banking and insurance services (Parenti, 2021). Although restrictions on capital flows were not allowed according to Article 67 of the ToR, several countries did not lift all barriers to capital flows until the end of the 1980s (Grahl, 2009). For instance, Belgium and the Netherlands were able to hinder capital flows with the purpose of keeping their domestic interest rates low. This was contrary to the intention of finishing the process laid down by the ToR by the end of 1969. However, it can be explained by the fact that the ToR did not entail many details on actual regulations, but rather was composed of general statements and objectives (Dedman, 2010). The restrictions of capital flow, the narrowness of capital markets, and the unilateral preference for domestic borrowers in Europe raised criticism from the United States, the OECD, and the Segré report (Kindleberger, 2015). As a result, the Council proposed several modifications of present regulations towards greater liberalisation which were implemented through several directives in the period between 1964 and 1973 (Parenti, 2021). The end of the fifties was also the time when the Eurodollar market developed where European banks conducted transactions in US dollars since they were not affected by Regulation Q which set interest rate limits for US American banks (Kindleberger, 2015). Most of these transactions took place in London. The size and importance of this market triggered a debate about the regulation of this market. In the beginning, the Bank of England only requested banks to report their Eurodollar holdings. Compared to continental European governments, the Bank of England did not shut down the market, since the goal was to establish London as a major financial capital. Furthermore, the capital inflows were also desirable to solve the United Kingdom's balance of payments problems. On the other hand, the United States also did not slow down the market. The situation was different for continental European countries as they were much more prone to capital controls. This applied even to more liberal countries like Germany which reintroduced controls on capital inflows in the sixties. As a response to the development of the Eurodollar market, many continental European countries imposed mandatory reserve holdings on Eurodollar investments and prohibited paying interest on foreign deposits. Although it was the goal to cooperate in regulations among European countries, progress was slow. The EEC installed a Contact Group of national supervisors in 1972, but it was only in 1977 that they came up with the first directive on banking (Schenk, 2010).

In 1985, the Commission published a White Paper called "Completing the Single Market" which envisaged the liberalisation of financial services. The Single European Act (1987) transformed the objectives of the White Paper into an EU law and had the intention to complete the internal market by 1992 (Parenti, 2021). This drew the attention of member states of the European Free Trade Area (EFTA) as they saw the great potential of the internal market – especially given the collapse of the socialist systems in Eastern Europe. This led to the establishment of the European Economic Area (EEA) which included EEC and EFTA countries. If everyone had agreed to the EEA, it would have been composed of 19 countries all applying the four freedoms. In the end, two EFTA states (Liechtenstein and Iceland) did not opt to join the EEA, and in two countries (Norway and Switzerland) EEA membership was rejected by referenda (Dedman, 2010).

In the end, capital markets in most European countries were only liberalised in 1990 (Dedman, 2010). This led to a narrowing of the interest rates gap in the member states where the interest rates in European countries had been mainly driven by the German market. The combination of free capital flows and tied exchange rates among European Monetary System (EMS) countries implied that active monetary policy with a focus on output stabilisation was not possible anymore. This made the European leaders think about the usefulness of a common currency for all EMS countries (Pittaluga, 2016). The introduction of a common currency was a way for the other European countries to regain more control over monetary policy and to be less dependent on German monetary policy decisions. Another idea behind the common currency was the experience of the financial turmoil of the late 1960s. The logic was that a common currency should make such financial turbulence less likely (Dedman, 2010). Finally, the joint currency was introduced by the Maastricht Treaty and it brought the capital markets closer together. Since 1999, all commercial banks from Euro member states can get liquidity from the same institution – the European Central Bank (ECB). As financial markets were still partly segmented in 1999, the Commission formulated 42 additional measures as part of a Financial Services Action Plan (FSAP) that was scheduled to be introduced by 2004 (Parenti, 2021). This FSAP helped to bring European commercial banks closer together and to intensify the competition among these banks. In this respect, the regulation contributed to a convergence of the profitability of Euro area banks (Goddard et al., 2013). The focus then went to other market segments. A committee of wise men which was introduced by ECOFIN and headed by Alexandre Lamfalussy published a report with suggestions on how to regulate securities markets in the EU. These recommendations – also called the "Lamfalussy process" – established a four-level approach for regulation (Parenti, 2021). The "Lamfalussy process" made EU legislation more transparent, and enhanced cooperation among different internal actors and external stakeholders which improved the quality and the speed of EU legislation (Schaub, 2005). An important part of the "Lamfalussy process" was delegation, which, in fact, made negotiations faster and the regulation of securities markets more effective (De Visscher et al., 2008). Clearly, the regulation of financial services in the EU is a topic on its own and goes far beyond what we discuss here. For instance, Quaglia (2010) discusses this topic in detail in his book.

After the first years of the Euro, the main event that changed everything was the Subprime Crisis originating in the United States. This major economic crisis sent shockwaves from the United States to the European banking system. Legislators all over the world tried to come up with responses to this financial crisis. One answer was the Basel III accord which included stronger requirements for liquidity and equity capital for banks. As the Basel III accord is based on discussions of the Basel Committee, including countries from all over the world, it has no binding character and must be implemented in law. Despite the fact that Basel II did not prevent the financial crisis of 2008, the process of implementing Basel III was complicated in the EU since some countries were reluctant to introduce the new capital requirements (Howarth & Quaglia, 2013). Apart from that, several other directives were implemented to strengthen banking regulations in the EU. However, authors like Quaglia (2013) question whether this will be enough to prevent further financial crises in the EU. After this, Europe was faced with its very own crisis – the so-called Euro crisis – which has many different origins (C. S. Weber, 2015). One of the answers of the EU was to introduce a banking union with the goal of supervising the largest and systematically relevant banks of every Euro country. The primary goals of the European Banking Union are to solve the doom-loop problem of mutual dependence on state and bank solvency. Furthermore, it attempts to find a solution for the financial trilemma of financial stability, cross-country banking, and national financial regulation (Schoenmaker, 2016).

It took a long time until the topic of sustainability was included in the debate about financial regulation. Ahlström and Monciardini (2022) identify three phases in the transformation of financial regulations towards sustainability. Until 2009, sustainable finance was not at all on the EU's agenda. The first phase in the direction towards more engagement in sustainable finance lasted from 2009 to 2012 and was conflictual. During this period, finance and sustainability were seen as incompatible, regulation was regarded as being counterproductive, and the focus was solely on profit maximisation. Sustainable finance was living a niche life. The subprime crisis and the Euro crisis changed the view of many people about the financial sector and regulation. It became apparent that a *"laissez-faire"* approach was no longer suitable and that EU citizens questioned the usefulness of the financial industry. In 2011, the Commission mentioned for the first time the desire to introduce a general rule for the provision of information about sustainable investments.

The second phase that Ahlström and Monciardini (2022) identified was from 2012 to 2017. During this period, the topic left its niche status and was discovered by larger financial enterprises and NGOs. This period saw the introduction of a high-level expert group on sustainable finance in 2016. A milestone was the Paris Agreement of 2015 which enlisted "making finance flows consistent with a pathway towards low greenhouse gas emissions and climate-resilient development" (UNFCCC, 2015) as one of the main objectives in Article 2c. The EU signed the Paris Agreement on behalf of its member states. The next important step was the 2017 revision of the EU Shareholder Rights Directive (SRD) – originally from 2008 – which included a section about the role of institutional investors (Ahlström & Monciardini, 2022). Article 3g of Chapter Ib of the SRD said that "institutional investors and asset managers shall develop and publicly disclose an engagement policy that describes how they integrate shareholder engagement in their investment strategy" (European Union, 2020). More narrowly, the focus of the EU's actions and reforms was mainly on climate finance and on fostering the establishment of a green finance market (Ahlström & Monciardini, 2022).

The third phase identified by Ahlström and Monciardini (2022) took place from 2017 to 2019, but we can extend this period until now. This period saw a continued focus on green finance with little emphasis on other subjects of ESG criteria. The main programme here was the Sustainable Finance Action Plan (SFAP) in 2018. The goal of this SFAP was to come up with an EU taxonomy, create standards for an EU Green Bond market, increase the transparency of investments, include climate risk in credit ratings, and stimulate sustainable investments (Parenti, 2021). However, important ESG aspects such as human rights and social aspects only played a minor role in EU taxation, which raised criticism from NGOs (Ahlström & Monciardini, 2022). Article 18 of the EU Taxonomy has some regulations on human rights and labour rights, but governance aspects are not mentioned at all (de Oliveira Neves, 2022). The EU taxonomy was implemented in 2020. A critical aspect was the role of gas and nuclear energy in EU taxonomy. Ultimately the Complementary Climate Delegated Act of 2022 states that nuclear energy and gas can be considered sustainable under certain conditions. The next step after the SFAP is the Corporate Sustainability Reporting Directive, which will make reporting and assessing sustainability risks obligatory (de Oliveira Neves, 2022). Notably, the regulation requires the reporting of all ESG aspects.

Innovation

The Important Role of Innovation in Economic Transition

Environmental issues and climate change are high on the agenda of politicians but also among citizens. According to a Special Eurobarometer 501 conducted before the start of the COVID-19 crisis, 96% of the French survey participants (94% among all participants) considered that protecting the environment is important for them (European Commission, 2020). The most frequently mentioned environmental issues were air pollution (50%), climate change (49%), and increasing amounts of waste (48%). In a recent survey (Allianz Research, 2021), an absolute majority of the respondents evaluated the 55% CO_2 reduction goal of the EU by 2030 as plausible or not ambitious enough. As the main measure to reach climate goals, the French respondents referred to "research for and funding of new sustainable technologies". If we combine this with the fact that 38% of the French respondents see economic growth as the best way to overcome public debt resulting from COVID-19 rescue measures, then this shows that a de-growth strategy might not be a desirable strategy.

If economic growth should persist but at the same environmental targets are to be reached, this puts a spotlight on the role of total factor productivity (TFP) and innovation. Improvements in TFP – for instance, through a reduction of input factors while maintaining the same output – can lead to economic growth and improvements in the environmental balance. Innovation plays an important role in this context – while being a prerequisite for increasing TFP[3] – and is seen as a driver for improving environmental sustainability (OECD, 2011; Silvestre & Țîrcă, 2019).[4] Mahlberg et al. (2011) can prove for a subset of 14 EU countries that the majority of productivity growth actually comes from energy saving.

One proxy for innovation is spending on R&D of companies of countries in general. There are several studies which use this approach. For instance, Pegkas et al. (2019) provide evidence for a sample of EU countries that R&D expenditure leads in fact to innovation. And then there is some support for the idea that innovation leads to a lower environmental footprint. Fernández et al. (2018) conducted an empirical study on the EU15 for the years 1990–2013. According to them, R&D significantly reduced CO_2 emissions in EU15 countries over that time span. Shahbaz et al. (2018) focused on just one EU country (France) and they demonstrated that real R&D expenditure for energy innovation helped reduce CO_2 emissions.

However, if TFP growth does not necessarily lead to an improved environmental balance, then this opens a debate about the relationship between innovation and sustainability. However, not all innovations are sustainability-enhancing and there is no certainty that productivity growth will be beneficial for the environment. Therefore, scholars and practitioners distinguish between innovation and eco-innovation (also called green innovation, sustainable innovation, or environmental innovation).

According to the empirical literature (e.g., Kesidou & Demirel, 2012; Triguero et al., 2013), there are three main drivers of green innovation: supply-side effects (technology push), demand-side effects (demand pull factors), environmental regulation, and the organisational culture of firms. In the context of European studies, the role of regulation is the most interesting as it directly tackles the role of European institutions. The impact of environmental regulation is obvious. If certain production techniques or input factors are forbidden, firms have no other option but to adjust their production processes or abandon the production of the respective products. What might be less obvious is that regulation can be productivity enhancing. At first glance, traditional economic theory might suggest that higher regulation – regardless of the aspect of business life – is harmful to business activity. However, this naive view was questioned by Michael Porter in his seminal work (Porter & Van der Linde, 1995) as he argued that environmental regulation can actually be innovation and productivity enhancing. Three versions of the Porter Hypothesis (PH) can be distinguished. The weak version of the PH posits that environmental regulations lead to innovation. The strong version goes one step further and hypothesises that environmental regulations lead to higher productivity. Furthermore, there is also a narrow version which says that the effects of regulation on innovation and productivity are greater if regulation is founded on market-based policies.

Several studies provide empirical evidence for the weak version of the PH. Triguero et al. (2013) conducted an empirical study among 5,222 managers of SMEs in the 27 EU member states at that time (2011). They show that existing environmental regulation is one of the drivers of eco-innovation. Triguero et al. (2013) found another way the public sector enforces green innovation, and they provide empirical evidence that networks with universities and research institutions significantly increase eco-innovation. In the empirical study of Horbach (2008), compliance with regulatory measures is one of the motivations for companies to eco-innovate. An indirect way of assessing the role of environmental regulation is to look at abatement costs. Kesidou and Demirel (2012) find a positive effect of abatement costs on green innovation. This confirms that environmental regulation fosters innovation, especially for low-innovative and high-innovative firms. In another study on 17 European countries, there is also evidence that Pollution Abatement and Control Expenditures (PACE) foster the creation of patents (Rubashkina et al., 2015). The main part of environmental regulation is environmental taxes which are the focus of the empirical study of Franco and Marin (2017) on eight European countries. Using data from the European Patent Office, they find that both upstream environmental taxes and downstream environmental taxes can encourage patents, whereas within-sector environmental taxes have no significant effect.

The evidence for the strong version of the PH is less pronounced. For instance, Rubashkina et al. (2015) prove the weak version of the PH but they do not find any evidence for the strong version of the PH which says that regulation enhances productivity. On the other hand, there are also some studies which argue that the strong version of the PH holds up. The introduction and the adoption of green technologies are considered a cost effective way to reduce environmental pressure without compromising economic competitiveness (Costantini et al., 2017). Among them is the one of Franco and Marin (2017) which demonstrates that within-sector environmental taxes and downstream environmental taxes both significantly increase the TFP. Costantini and Mazzanti (2012) analyse the EU15 and lay their focus on the trade effects of regulations. According to their empirical study, environmental taxes either increase export dynamics, or have no effect. When looking only at environmental goods, there is a robust positive effect of environmental taxes on export dynamics. The same can be said about PACE. Rexhäuser and Rammer (2014) demonstrate with a large sample of German companies (29,807 enterprises) that both forced and voluntary resource efficiency-improving innovation increase the returns on sales of the respective companies.

The Important Role of Innovation in Sustainability Transitions

The recognition of many environmental problems, such as climate change, loss of biodiversity, and resource depletion, is one of the main motivations for research on sustainability transition (Elzen et al., 2004). Research on sustainability transition aims to characterise how radical changes can occur. This research focuses on the meso level of socio-technical systems (Geels, 2004). Sustainability transition studies rely on different theoretical frameworks such as the multi-level perspective (MLP), the technological innovation system (TIS) approach, strategic niche management (SNM), and transition management (TM) (Smith et al., 2010). These theoretical approaches combine ideas from evolutionary economics, in particular innovation systems research, the sociology of innovation, and institutional theory.

Every country should ultimately reach a low-emission development path because it yields several benefits beyond climate change (Trærup, 2022). The energy sector seems to be a key focus for reducing greenhouse gas emissions (GHG). Indeed, renewable energy sources can free countries from dependence on imported fossil fuels. Therefore, switching to a low-carbon economy has the potential to yield substantial economic benefits. By implementing energy-efficient measures, we can increase our income levels. In addition, low-carbon technologies can introduce new sources of growth and employment opportunities. Furthermore, the development and adoption of innovative technologies could provide some of the world's poorest countries with a comparative advantage.

Since its establishment in 2001 as a part of the United Nations Framework Convention on Climate Change (UNFCCC), the technology needs assessments (TNAs) have been developed specifically to conduct comprehensive analyses. These assessments are described as "a series of initiatives led by individual countries to identify and prioritise the technologies required to mitigate and adapt to climate change." When it comes to developing frameworks for the widespread use of climate technologies, it is crucial to consider the cost and accessibility of financing. Based on what countries have reported through their TNAs, it is evident that access to capital and investments remains a significant challenge for developing and vulnerable nations seeking to obtain and utilise green technologies. The TNAs have identified specific challenges faced by these countries in developing, using, and transferring green technologies (Trærup, 2022).

A Quick History of the Innovation Regulation in the EU

The EU's objective to foster "scientific and technological advance" is laid down in Article 3 of the Treaty on European Union (TEU) as one of the main objectives of the EU. This goal is mentioned right after the aim to achieve "sustainable development of Europe", which shows that the EU does not see sustainability and technological progress as rival goals (Blanke & Mangiameli, 2013).

This provides the overarching mission, while the details are regulated by the Treaty on the Functioning of the European Union (TFEU). The legal basis for innovation policy is enshrined in Article 173 TFEU. This article states that the objective of industrial policy is – among other things – to foster "better exploitation of the industrial potential of policies of innovation, research and technological development". Initially, the Commission wanted to harmonise industrial policy back in 1973, but it took much longer until the MS agreed to add this to EU regulations. Notable initiatives based on this article are the European Fund for Strategic Investments and the European Institute of Innovation and Technology (EIT) (Kellerbauer et al., 2019). The EIT was introduced in 2008 and made use of Knowledge and Innovation Communities with actors from the business, research, and education sectors (Gouardères, 2021). The European Fund for Strategic Investments and the European Investment Fund, and the European Investment Bank provide funding for small and medium enterprises with the aim of fostering venture capital investments, strategic investments, and investments in sustainable growth (Gouardères, 2021). The follow-up programme InvestEU provides funding during the period 2021–2027 with a budget of around 370 billion euros. InvestEU is an integral part of the European Green Deal (d'Alfonso, 2020).

In addition, Article 179 TFEU implements the European Research Area which has its origin in an initiative of the Commission in 2000 (Kellerbauer et al., 2019). The idea was that Europe should catch up with the United States and improve its competitiveness with an economy based on knowledge and information (Craig & De Búrca, 2020). Furthermore, the European Research Council was created in 2007 with a substantial budget to finance research activities (Kellerbauer et al., 2019). With Horizon 2020, the EU implemented a large research and innovation programme with a budget of around 80 billion euros, and the succeeding programme, Horizon Europe, has a budget of around 95 billion euros.

In the EU, innovation policies are incorporated through different measures. It starts with the protection of innovation through patents and trademarks. The European Union Intellectual Property Office (EUIPO) handles trademarks and design rights within the EU. The EUIPO was established in 1994. Dealing with patents lies in the hands of the European Patent Office (EPO). Patents are possible at the national level or the EU level. If someone wants EU-wide protection of an idea, then the person must register their innovation with the EPO and apply for a European patent. However, in this case, it is also necessary to let the patent get acknowledged in each country where the person wants protection of the idea (Fabrizi et al., 2018; Sfârlog, 2020; Ullrich, 2002). The EPO was established in 1977 through the European Patent Convention (1973). It is not part of the EU although all EU MS are also member states of the EPO. In addition, several non-EU countries hold a membership in the EPO. However, EU law – notably Article 36 of the TFEU – guarantees that the national procedures for protecting intellectual property rights and patents are applied to any EU citizen in the same way (Fairhurst, 2016). The establishment of the EPO substantially changed the administrative effort for patents. The private value of patents by the EPO is higher than those given by individual member states (Deng, 2007).

An indirect way to foster eco-innovation is the labelling of products. The Energy Label of the EU which was introduced in 1994 is well known among customers and helps them make more eco-friendly decisions. The original Energy Label was applied to a larger number of products in 2004 and underwent a major reform in 2021. The reform of 2021 is the result of the success of the previous Energy Label as most fridges and washing machines were labelled A+ or better. As energy efficiency increased so much over time, the new Energy Label requires even higher energy efficiency (Hinchliffe & Akkerman, 2017; Talens Peiró et al., 2020). Empirical research shows that the Energy Label in combination with minimum energy performance standards contributed to an increase in the market share of products rated A+ or higher of around 15–38% (Schleich et al., 2021). However, these labels can be controversial, as can be seen in the case of Dyson vacuum cleaners. Dyson sued the EU saying that the energy labels are misleading. He won the trial at the General Court, but he lost his fight for compensation. The result is that vacuum cleaners no longer have to be labelled in the EU (Chee, 2021; Koolen, 2019).

In the mid-1980s, the European Framework Programmes (FP) for research and innovation were established. According to Kastrinos and Weber (2020), the term "sustainable development" appears for the first time in the EU Research & Innovation vocabulary in the Fourth Framework Programme (1994–1998). The Fourth Framework Programme addressed sustainable economic development and sustainable economic growth.

Another important aspect is that the EU Commission provides a European Innovation Scoreboard (EIS). The EIS was first provided in 2006 and is published every year. It provides innovation scores for all individual EU member states. In addition, there is also a Regionalised Innovation Scoreboard (RegIS) which provides innovation scores on a regional level. The EIS has 25 dimensions in total with five main categories focusing on both innovation output and application (Iking, 2009). Both EIS and RegIS help to compare the innovation levels of countries and regions which can serve as an important basis for policymaking. According to Korres (2013), we can distinguish between two main dimensions of innovation: social innovation and technical innovation. The EIS includes both of these dimensions. A main tool to foster innovation is the EU's Europe Strategy 2020 which is the general growth strategy of the EU. The EU recognises that economic growth can be achieved through smart growth, sustainable growth, and inclusive growth. Apart from that, the EU has a European regional innovation strategy (Korres, 2013). In addition, there is also the EIT Regional Innovation Scheme, introduced in 2014, aims to support regions with low innovation scores (*EIT* Regional Innovation Scheme, 2019).

Two main changes were introduced with the Horizon 2020 programme: missions and the European Innovation Council (EIC). The EIC was launched in

2018, but the idea to establish an EIC dates back to June 2015 with a speech by Carlos Moedas, Commissioner for Research, Science and Innovation (M. Weber et al., 2019).

External Relations

The EU could play a major role in addressing sustainable development challenges. The role of the EU is manifold: it is both a consumer and a producer, it is also a funder as well as a rule maker and an enforcer (Hedberg & Šipka, 2022). The EU could use the power of legislation (directives, regulations), economic instruments (public funds), and soft law (e.g., guidelines, voluntary commitments) to reach sustainable development. Therefore, the role of the EU and its external relations are important to address environmental challenges.

The following summary of external relations at the EU level has been chosen for its connections between finance, innovations, and environmental policies.

The Influence of External Relations on Environmental Transition in the EU: General Agreements about Climate Change and Environmental Issues

The EU has implemented several agreements across the world to promote sustainability. The existence of these agreements underlines the importance of the EU's external relations to achieve sustainability. These agreements deal with environmental as well as climate change issues. Environmental policies encompass a wide range of issues such as air, water and soil quality, biodiversity conservation, waste management, and sustainable development. Climate change policy is more specific and deals with the accumulation of greenhouse gases in the atmosphere and the need to reduce their concentration.

At the EU level, several measures have been adopted to tackle climate change issues. One of the most well known measures was the EU Emissions Trading Systems (ETS), which was introduced through the Kyoto Protocol adopted in 1997 and enforced in 2005. The ETS limits the amount of greenhouse gas (GHG) emissions that some industries can produce and authorises firms to buy and sell emissions permits. Indeed, firms that emit less than their allocation can sell their surplus on the market: they are financially incentivised to reduce their GHG emissions.

Besides, from a historical perspective, the EU has been involved in the Intergovernmental Panel on Climate Change (IPCC) since its creation in 1988 by the United Nations. The EU was also one of the original signatories to the UNFCCC. More recently, the EU is still actively engaged in international efforts to address climate change with the development of the Paris Agreement in 2015 and the European Green Deal in 2019. The latest aimed to make Europe climate-neutral by 2050 to achieve net-zero GHG emissions by 2050 in line with the Paris Agreement's temperature objective. It also aimed to propose a law on climate neutrality (European Climate Law).

Regarding environmental issues, the Lisbon Treaty in 2009 recognised the need to achieve sustainable development at the EU level. It provided the EU with new powers to address environmental issues including the ability to pass binding climate change agreements as well as binding agreements to promote global environmental sustainability (Benson & Adelle, 2012).

Other measures include the development of renewable energy sources and the development of low-carbon technologies through R&D programmes such as the Horizon 2020 programmes. Regarding the development of renewable energy sources, the EU has set a target of reaching 32% of its final energy consumption from renewable sources by 2030. To achieve this target, the EU has implemented the Renewable Energy Directive in 2018 (Krug et al., 2022). Besides, the EU has developed numerous research and innovation programmes. For example, Horizon 2020 aims to support the development of innovative technologies that can reduce GHG emissions. This programme covered the period 2014–2020 and it is considered as one of the main research and innovation programme with a budget of €80 billion. This programme has also funded research about the impact of climate change and the policy measures that could be implemented to address this impact (Pollex & Lenschow, 2018). Horizon Europe has succeeded Horizon 2020. It covers the period 2021–2027 with a budget slightly higher than Horizon 2020 (Latoszek, 2021; Ricciardiello et al., 2021).

Furthermore, the EU has been active in negotiating agreements to protect biodiversity and promote the conservation of endangered species and their habitats. One can mention, for example, the EU-Africa Wildlife Law Enforcement Dialogue which was launched in 2015. It is a platform to establish cooperation between the EU and African countries on the issue of wildlife trafficking. This platform includes the European Commission, the African Union Commission and the Secretariat of the Convention on International Trade in Endangered Species of Wild Fauna and Flora (CITES). This is a multilateral agreement which aims to regulate the international trade of wild animals (Haastrup et al., 2021). To the best of our knowledge, attempts to discuss the impact and consequences of sustainability policy on EU external relations are scarce. The book by Adelle et al. (2018) deals with the environmental policies (forests, biodiversity, water, climate change) that the EU employs outside its borders (Africa, China, USA, Latin America). Our contributors (three chapters respectively written by Lucie Qian Xia, Frauke Pipart, and Paiman Ahmad, Muhammad Usman and Atif Jahanger) aim to analyse EU-style diplomacy, its impact on environmental negotiations, and the European Green Deal effect on oil-producing economies which seems not to have been studied yet.

Agreements Related to Finance, Innovation, and Sustainability

We provide more detail in the following about the agreements related to finance, innovation, and sustainability.

The EU's Digital Single Market Strategy (DSM) serves as a key example. The European Commission launched the Digital Single Market Strategy (DSM) in 2015 in order to create a single, unified market for digital goods and services across the EU. The DSM aims to reduce cross-border online trade barriers and promote the development of the EU's digital economy. The strategy is built on three main pillars: improving access to digital goods and services, fostering the growth of digital networks and services, and promoting the digital economy's growth. The DSM includes various financial aspects such as funding programmes to support the development of high-speed broadband networks (some of which are financed with the support of the Horizon 2020 programme) financial support for SMEs in the digital sector (through the European Digital Innovation Hubs). The DSM also includes policies that aim to promote a more environmentally friendly digital economy in the EU. It comprises the objective of measuring the carbon emissions associated with digital technologies such as the data centres (Energy Efficiency Directive and the Green Public Procurement Initiative) and mobile network, reducing the carbon footprint of digital technologies, promoting the circular economy by reducing electronic waste and reusing digital devices and components (Circular Economy Package), and encouraging the use of digital technologies that can support sustainability such as smart grids (Alvarez León, 2018). The DSM is an example of policies which address cross-border challenges related to digital technologies such as online privacy and data protection (with the EU–US Privacy Shield and the EU–Japan Economic Partnership Agreement), cybersecurity, and intellectual property rights.

The EU–Australia Strategic Partnership Agreement (SPA) is a comprehensive bilateral agreement between the EU and Australia. This agreement was signed in 2018 and entered into force in 2019. It aims to deepen the cooperation between the EU and Australia in several areas: trade and investment, foreign and security policy, and research and innovation. This agreement includes commitments to increase trade and investment flow between the EU and Australia. More precisely, the agreement aims to reduce barriers to trade and investment. It also provides for the protection of intellectual property rights which in turn favours innovation and investment in new technologies. Besides, the SPA also encourages sustainable development by promoting investment in green technologies and reducing the environmental impact of economic activities (Murray, 2016). The SPA is another example of an agreement for which financial, sustainable issues, and external relations are at stake.

The EU–China Comprehensive Agreement on Investment (CAI). It is a trade agreement which was negotiated between the EU and China. The negotiations started in 2014 and were concluded at the end of the year 2020. This agreement aims to improve EU companies' access to the Chinese market (in the sectors of the automotive, financial services, healthcare, and the telecommunications for instance) while also encouraging a fair competition between the EU and China. This agreement addressed other issues such as investment protection, long-term development, and sustainable development. Commitments to promote responsible business conduct and corporate social responsibility are included in the CAI such as the protection of labour rights including the eradication of forced labour, child labour, and the rights of collective bargaining and the freedom of association (Abgaryan, 2018). For the environmental part, the CAI includes commitments to combat climate change and reduce GHG emissions in line with their objective under the Paris Agreement. The CAI forms a Sustainable Development Working Group to monitor implementation and dispute settlement arrangements (Cotula, 2021). This can lead to new opportunities for innovation and cooperation. For example, it could create opportunities in areas like renewable energy, electric vehicles, and smart cities, which in the end can help to foster a more sustainable approach to economic growth and investment between the two sides. The CAI has also significant implications for innovation and technological cooperation between the EU and China. The CAI aims to make it easier for EU companies to invest in China's innovation sector. This is possible because the agreement commits to reducing or eliminating restrictions on foreign investment in various areas such as telecommunications, cloud computing, and R&D. This will allow EU companies to invest in China's thriving innovation ecosystem that is home to many fast-growing technology startups and research institutions. Moreover, the CAI also has provisions to protect intellectual property rights and tackle counterfeit and piracy issues. This could benefit EU companies as it provides greater protection for their intellectual property in China. Therefore, the agreement fosters greater innovation and technological cooperation between the EU and China. However, we should mention that this agreement is criticised regarding the treatment of the Uighur minority in China's Xinjiang region (Casarini & Otero-Iglesias, 2022).

A Quick History of the External Relations in the EU

The ECSC marked the beginning of the process of the European integration in 1951. The objective of the ECSC was to promote economic growth and prevent war by pooling coal and steel production in Europe. The ECSC can be considered as the starting point of the EU's external relations policy. Then, in 1957 the EEC was created by the ToR to pursue further European integration by establishing a common market for goods, services, capital, and labour. The ToR is considered as an important milestone in the development of the EU's external policies because the EEC allowed a unified approach to trade negotiations with non-EEC states. Indeed, prior to the EEC's establishment, individual member states negotiated trade agreements independently, which often led to conflicting policies. The EEC was empowered to represent its member states in international trade negotiations. For example, the EEC member states agreed to a reduction of their external tariffs and trade barriers during the Kennedy round (1964–1967) and the Tokyo round (1973–1979) in the framework of the General Agreement on Tariffs and Trade (Dür, 2008).

The Maastricht Treaty, which set the EU, was signed in 1992. This Treaty introduced a distinction between the intergovernmental methods and the community method (Schäfer, 2006). The intergovernmental method empha-sised the role of national governments in decision-making. In this approach, decisions are made unanimously by the member states. For this reason, this method is considered as reflecting the interests of national governments. In contrast, the role of EU institutions in decision-making is accentuated in the community method. This method enables a greater participation of EU insti-tutions which have independent powers from member states. Indeed, before the Maastricht Treaty, cooperation in foreign policies, security, justice, and inter-nal affairs were the responsibility solely of the countries themselves (Duff et al., 2002). The Maastricht Treaty introduced two main policy areas: the Justice and Home Affairs and the Common Foreign and Security Policy.

The Justice and Home Affairs implemented a common approach to justice, law enforcement, and other issues related to internal security such as the fight against crime and terrorism, migration policies, and civil protection. The Justice and Home Affairs also aims to ensure that EU's policies are consistent with the human rights (Christiansen et al., 2012; Monar, 2004). Even if the Justice and Home Affairs is primarily focused on internal EU matters, the domains that are concerned by the Justice and Home Affairs have some impacts and consequences outside the EU. For example, the EU's migration policies have significant impacts outside the EU and require cross-border cooperation (Finotelli, 2018)

Besides, by establishing the Common Foreign and Security Policy (CFSP) the Maastricht Treaty has enabled EU member states to develop a common direction of the foreign policies. More precisely, the CFSP aims to maintain peace and international security. However, the CFSP is sometimes criticised for having a slow decision-making process in situations of international crises such as the ones that took place with the Gulf wars (1990–1992), as well as in Bosnia (1992–1995), and in Kosovo (1998–1999) (Knodt, 2003). More recently, the EU's response to the ongoing conflict in Ukraine has also faced criticism for being slow and ineffective (Filipec, 2017).

In 2009, the Lisbon Treaty introduced numerous changes to the EU's external relations policy (Duke, 2008). First, the position of the High Representative of the Union for Foreign Affairs and Security Policy was extended to include the role of Vice-President of the European Commission. Currently, Josep Borrell Fontelles holds this position. He is responsible for coordinating the EU's external relations and foreign policy. Second, the European External Action Service (EEAS) was established by the Lisbon Treaty. It is the primary institution in charge of manag-ing the EU's external relations in this field. The EEAS plays a key role in strength-ening the EU's relationships with its partners around the world and in addressing climate change and sustainable development issues. Indeed, the EU has also been working to deepen its relationship with other regions of the world such as Africa (with the Africa–Europe Alliance for Sustainable Investments and Jobs) and Asia (through the EU–Asia Connectivity Strategy). We could also notice several major trade agreements such as the Transatlantic Trade and Investment Partnership with the United States, the Economic Partnership Agreement with Japan, and the Comprehensive Economic Trade Agreement with Canada.

Therefore, since the Maastricht Treaty, the EU's policy scope has been significantly extended to include a wide range of issues such as finance, innovation, agriculture, transport, and regional development. Besides, the Lisbon Treaty is generally considered as having strengthened the EU as an international actor even if in the domain of foreign policies the willingness to cooperate with the member states is still crucial (Koehler, 2010). The history of the EU's external relations has been marked by a growing recognition of the EU's role as a global actor regarding economic issues as well as environmental issues such as climate change and sustainable development.

Theoretical Approach

Analysing the integration in any policy field including finance, innovation, and external relations requires a theoretical framework. There is now a long history of environmental regulation in the EU. Environmental problems became a public topic in the 1960s, leading to the recognition of the "tragedy of the commons". The result was the first UN conference in the 1970s (Dyer, 2017). This is also the time when environmental regulation was introduced in the EU, even though it was not explicitly mentioned in the Treaty. This can be attributed to the pressure of interest groups and opportunism by the Commission (Laffan & Mazey, 2006). The Commission developed a Community agenda for fields like environmental policy where there was no clear regulation in the treaties. It was the Single European Act which gave the European communities some competencies for environmental policy along with the four freedoms (Christiansen, 2006). The Single European Act introduced horizontal competencies like R&D and environmental protection. In the case of financial services, the competencies are vertical (Thatcher, 2006). The next milestone was the TEU which gave more competencies to the Community in the field of environmental policy and gave the European Parliament veto rights for environmental policies (Laffan & Mazey, 2006). Furthermore, the Maastricht Treaty introduced a Qualified Majority Voting (QMV) for environmental policy. The Treaty of Amsterdam and the Treaty of Nice further expanded the role of the EU in environmental policy (Thatcher, 2006). Given these facts, we can conclude that there has been substantial integration in the fields of environmental policy in the EU during the last decades. That being said, environmental policy in the EU was and is highly pluralistic (Richardson, 2006).

Typically, studies employ traditional integration theories like neofunctionalism, intergovernmentalism, or liberalism. There are several arguments as to why these traditional theories might not be suitable as a framework for our book. We will quickly discuss the pros and cons of different integration theories before we explain the decision for the theoretical framework for the book.

Let us start with realism: for realists, environmental problems are another point on the list of possible conflicts between states (Sørensen et al., 2022). However, international cooperation with the objective to address climate change might contrast with national interests (Theys, 2017). Considering that the climate crisis is an international crisis, it requires a global solution. The political answer to this problem could be, however, either a centralised system or a de-centralised but very interconnected system. Decentralisation would mean the inclusion of the local population in decision-making and a high level of democratic legislation. However, there are several problems related to this. First, a problem could be parochialism, which involves limited cooperation among local communities (Dyer, 2017). Second, there are some problems, like air pollution, which cannot be solved on a national basis. The solution to the problem of CO_2 emissions requires international cooperation and, to some extent, concessions by countries even though it could hurt their economies in the short run. Such cooperation cannot be explained by realism, which makes this theory unsuitable for the analysis.

The next in line is liberalism. To some extent, liberal internationalism (LI) could help explain integration in the field of environmental policy. The environment is the third major issue in world politics after security and the economy. As environmental issues can lead to international conflicts, it is just another argument for international cooperation for proponents of LI (Sørensen et al., 2022). Liberalism incorporates the idea that there is international law and international organisations which could serve specific objectives like addressing climate change (Meiser, 2022). However, there are four main reasons why liberalism can have difficulties to explain further integration in terms of environmental regulation. First, at the centre of liberalism is the idea of liberty, freedom of choice, and consumption. However, the consequences of these choices are ignored (Dyer, 2017). Second, liberalism has the issue that they fear regulation and favour economic growth (Theys, 2017). Third, LI acknowledges that public goods policies such as environmental regulation are more likely to be European-wide topics which can mobilise the engagement of society according to Moravcsik. However, he doubts that the political pressure will be effective, since voters care more about other policies, and the role of national parliaments is limited anyway (Leuffen et al., 2022). According to Moravcsik, the EU should focus on its business by improving market efficiency and correcting market failures (Bellamy & Attucci, 2009). Fourth, there could be a role of public opinion in LI theory as it has an impact on the interests of national leaders. However, it is unclear how strong the effect of public opinion could be (Kuhn, 2019). What we have seen is that the push towards more environmental regulation is very much driven by NGOs and public pressure on political authorities. Thus, we conclude that LI is not suitable as a framework for our book.

This brings us to neo-functionalism and post-functionalism. A neo-functionalist argument can explain why stakeholders and interest groups have pushed for more supranational regulation in a field like environmental policy. One example is that there are European-wide standards, for instance, for environmental regulation. This was in the interest of companies to have standardised regulation across Europe (Thatcher, 2006). Neo-functionalism proposes that there is a spillover from one field to the other (Jones et al., 2016). The problem is that neo-functionalism does not take public opinion into account. Politicisation is expected to promote, not to hinder, integration (Kuhn, 2019). As an alternative, post-functionalism emerged as a theory of integration. Hooghe and Marks (2009) are advocates of post-functionalist theory as an alternative because, in this theory, national leaders have to take public opinion into account. This is enforced by larger public debates due to more democracy within the EU (for instance through the European Parliament). Post-functionalism also acknowledges that the impact of political milieus was shrinking, starting in the 1980s, and moved more towards issues like the environment (Leuffen et al., 2022). This is an important aspect. However, post-functionalism cannot explain the reaction to the Euro crisis (Börzel & Risse, 2018). This is a problem, since greater integration in banking regulation and supervision was one of the milestones after the Euro crisis that has had a major impact on the European financial sector. Our conclusion is also in line with authors like Hooghe and Marks (2009) who argue that traditional integration theories are no longer relevant to European integration. They show that neo-functionalism and intergovernmentalism lose their impact on European integration.

As could be observed in the last Eurobarometer surveys, there is a growing recognition of environmental concerns (Bacsi, 2020; Baiardi & Morana, 2021; Tendero, 2021), especially among young people who indicate in these surveys that "protecting the environment and climate" is the second most important duty in the EU (Kuenzer, 2021). In these studies, environmental conscience is defined from a psychological perspective. In psychology, the cognition, the affect, and the conation are considered as important elements that allow a person to become aware of a phenomenon (Tallon, 2020). Cognition involves the acquisition and processing of information to rationally comprehend a phenomenon. Affect refers to the emotional responses triggered by this knowledge acquisition. Conation is the deliberate, intentional behaviour resulting from cognitive and affective experiences. A similar reading framework has been proposed by Sánchez and Lafuente (2010). They consider four main dimensions to define environmental consciousness: affective, cognitive, active, and dispositional (Jiménez Sánchez & Lafuente, 2010). The affective dimension corresponds to environmental concerns and the perceived severity of importance of environmental issues. This dimension is mainly reflected in the surveys investigating the attitudes towards environmental issues such as the Eurobarometers. The cognitive dimension refers to the level of information and knowledge of individuals regarding environmental issues (e.g., climate change, biodiversity loss, air, soil, and water contamination). The definition of this dimension is similar to the framework used by Tallon (2020). The active dimension corresponds to pro-environmental behaviours adopted by individuals. The dispositional dimension is linked to personal moral norms. It refers to the individual attitudes towards pro-environmental behaviours. Kollmuss and Agyeman (2002) provide a more comprehensive definition of environmental consciousness which includes factors such as environmental knowledge, values, attitudes, and emotional involvement. They also take into account the economic, social, and cultural context (Kollmuss & Agyeman, 2002). For example, the environmental attitudes in the EU are also related to per capita income (Scruggs & Benegal, 2012). This phenomenon is called the climate change/environmental awareness curve (Baiardi & Morana, 2021). As living standards and income are important factors to explain the existence of an environmental conscience, it is important to analyse the financial aspects of the sustainable transitions. This is even more important as the worldwide economic recession caused by the COVID-19 pandemic may negatively impact the environmental conscience. To sum up, from a psychological perspective, environmental consciousness could be considered a mental state composed of both personal and behavioural elements (Sharma & Bansal, 2013).

However, merely stating that the environment and climate change are significant concerns is not sufficient to demonstrate the existence and the importance of the European environmental conscience. It is in fact a necessary but not a sufficient condition. That is the reason why we can find in the literature different approaches to the concept of the environmental conscience. In political sciences, the environmental conscience includes the study of national and international environmental policies (in the broad sense: that is to say including climate change policies), laws, and treaties aiming at protecting the environment. It also includes the study of social movements and lobbying groups which advocate for a significant political change regarding the environment. These political and historical approaches that analyse policies and actions that have been undertaken over time to address environmental issues correspond to the framework developed by Hörber (2012) in his book *The Origins of Energy and Environmental Policy in Europe – The Beginnings of a European Environmental Conscience* (Hörber, 2012). In this book, the European environmental conscience is considered as a by-product of the European search for energy and safety (Hörber, 2012; Kuenzer, 2021). The European Green Deal reflects the growing environmental conscience among Europeans in terms of environmental policies. The book of Hörber and Weber (2022) has investigated using a historical perspective the development of the European environmental conscience. The authors have shown that the European environmental conscience has evolved over time, moving from a marginal and energy-centred concern to a central ideology in European politics. This work was done through an analysis of the results of the Eurobarometer surveys from 1974 to 2020 (Tendero, 2021) and through an analysis of the environmental policies over time.

Thus, we have opted to use a non-standard theoretical approach to the European environmental conscience (Hörber & Weber, 2022).

This is a theoretical framework that is flexible enough to adapt to different policy areas. Furthermore, it is convincing as it can explain why we see a trend towards more focus on environmental aspects in various policy areas. It also has the advantage that it gives relevance to the role of the public, which plays a major role in explaining the changes in regulation over the last years.

In order to identify whether specific developments regarding the policies and actions of the EU are part of a more general European environmental conscience, Hörber and Weber (2022) established the following four selection criteria:

- The connection between energy and environmental policies
- The innovative nature of solutions, or a commitment to progress
- European Commission expertise and therefore legitimacy for political leadership
- Grassroots support and demand for effective solutions

The authors of the individual book chapters will evaluate whether their specific cases meet all four criteria mentioned above. If all these points apply to the selected case studies, this would be further proof that European sustainabilism is a major driver for European integration as it can serve as a common mindset among Europeans. The following case studies will analyse in detail whether a European environmental conscience was at the heart of integration in each respective policy field.

The chosen methodology has already been applied in the edited book of Hörber and Weber (2022) with a different thematic focus. While their book focuses on energy and environmental policies, our book will examine other policy areas. The two main areas that our book refers to are (green) finance and innovation policies. This is very appealing as these policy areas are not originally linked to debates about sustainability, but instead, they show the impact that the growing concerns about the environment have on other policy fields. The underlying theoretical framework will serve as glue by sticking the individual chapters together. From this point onwards, the authors of the chapters will examine how well a genuine European environmental conscience can explain developments in European integration.

Notes

1 Despite this fact, many of the European environmental movement organisations are originally national movements which build alliances with organisations from other European countries (Rootes, 2002).
2 Power-to-liquid is a substitute for conventional fuels like kerosine. Together with biomass-to-liquid, they play an important role for reaching carbon neutrality in the EU (Blanco et al., 2018).
3 Adetutu et al. (2015) comes to a different conclusion. In their empirical analysis on EU countries, they show that a reduction of TFP is necessary if sulphur oxide (SO_2) and nitrogen oxide (NO_2) emissions should be reduced.
4 This conjecture is questioned by the empirical study by Carrión-Flores and Innes (2010) as they find that environmental innovations are not a main driver of reductions in emissions even though environmental policy targets foster innovation.

Bibliography

Abgaryan, S. (2018). EU-China comprehensive agreement on investment in the context of Chinese bilateral investment treaty program with the EU countries. *Indian Journal of International Law, 58*(1–2), 171–203. https://doi.org/10.1007/s40901-019-00095-8

Adelle, C., Biedenkopf, K., & Torney, D. (2018). *European Union external environmental policy: Rules, regulation and governance beyond borders*. Palgrave Macmillan.

Adetutu, M., Glass, A. J., Kenjegalieva, K., & Sickles, R. C. (2015). The effects of efficiency and TFP growth on pollution in Europe: A multistage spatial analysis. *Journal of Productivity Analysis, 43*(3), 307–326.

Ahlström, H., & Monciardini, D. (2022). The regulatory dynamics of sustainable finance: Paradoxical success and limitations of EU reforms. *Journal of Business Ethics, 177*(1), 193–212.

Allianz Research. (2021). *Allianz Pulse 2021: Old beliefs die hard* (p. 30). Allianz.

Alvarez León, L. F. (2018). A blueprint for market construction? Spatial data infrastructure(s), interoperability, and the EU Digital Single Market. *Geoforum, 92*, 45–57. https://doi.org/10.1016/j.geoforum.2018.03.013

Bacsi, Z. (2020). Environmental Awareness in Different European Cultures. *Visegrad Journal on Bioeconomy and Sustainable Development, 9*(2), 47–54. https://doi.org/10.2478/vjbsd-2020-0010

Baiardi, D., & Morana, C. (2021). Climate change awareness: Empirical evidence for the European Union. *Energy Economics, 96*, 105163. https://doi.org/10.1016/j.eneco.2021.105163

Barnes, P. M., & Hörber, T. C. (2013). Linking the discourse on sustainability and governance. In P. M. Barnes & T. C. Hörber, *Sustainable development and governance in Europe: The evolution of the discourse on sustainability* (pp. 19–33). Routledge.

Bellamy, R., & Attucci, C. (2009). Normative theory and the EU: Between contract and community. In A. Wiener & T. Diez, *European integration theory* (2nd ed., pp. 198–220). Oxford University Press.

Benson, D., & Adelle, C. (2012). EU environmental policy after the Lisbon Treaty. In A. Jordan & C. Adelle (Eds.), *Environment policy in the EU: Actors, institutions and processes* (3rd ed.). Routledge.

Blanco, H., Nijs, W., Ruf, J., & Faaij, A. (2018). Potential for hydrogen and Power-to-Liquid in a low-carbon EU energy system using cost optimization. *Applied Energy, 232*, 617–639.

Blanke, H.-J., & Mangiameli, S. (2013). The Treaty on European Union (TEU). *A commentary*. Springer.

Börzel, T. A., & Risse, T. (2018). From the euro to the Schengen crises: European integration theories, politicization, and identity politics. *Journal of European Public Policy, 25*(1), 83–108.

Carrión-Flores, C. E., & Innes, R. (2010). Environmental innovation and environmental performance. *Journal of Environmental Economics and Management, 59*(1), 27–42.

Casarini, N., & Otero-Iglesias, M. (2022). Assessing the pros and cons of the EU-China comprehensive agreement on investment: An introduction to the special issue. *Asia Europe Journal, 20*(1), 1–7. https://doi.org/10.1007/s10308-021-00641-3

Chee, F. Y. (2021, December 8). Dyson loses fight for $198 mln compensation over EU energy labelling rules. *Reuters*. https://www.reuters.com/business/dyson-loses-fight-198-mln-compensation-over-eu-energy-labelling-rules-2021-12-08/#:~:text=Dyson%20loses%20fight%20for%20%24198%20mln%20compensation%20over%20EU%20energy%20labelling%20rules,-By%20Foo%20Yun&text=BRUSSELS%2C%20Dec%208%20(Reuters),to%20EU%20energy%20labelling%20rules

Christiansen, T. (2006). The European Commission: The European executive between continuity and change. In J. Richardson (Ed.), *European Union: Power and policy-making* (pp. 96–118). Routledge.

Christiansen, T., Duke, S., & Kirchner, E. (2012). Understanding and Assessing the Maastricht Treaty. *Journal of European Integration, 34*(7), 685–698. https://doi.org/10.1080/07036337.2012.726009

Costantini, V., Crespi, F., Marin, G., & Paglialunga, E. (2017). Eco-innovation, sustainable supply chains and environmental performance in European industries. *Journal of Cleaner Production, 155*, 141–154. https://doi.org/10.1016/j.jclepro.2016.09.038

Costantini, V., & Mazzanti, M. (2012). On the green and innovative side of trade competitiveness? The impact of environmental policies and innovation on EU exports. *Research Policy, 41*(1), 132–153.

Cotula, L. (2021). EU–China comprehensive agreement on investment: An appraisal of its sustainable development section. *Business and Human Rights Journal, 6*(2), 360–367. https://doi.org/10.1017/bhj.2021.16

Cozzi, G. (2016). Finance and investment in the Eurozone. In G. Cozzi, S. Newman, & J. Toporowski (Eds.), *Finance and industrial policy: Beyond financial regulation in Europe* (p. 167). Oxford University Press.

Craig, P., & De Búrca, G. (2020). *EU law: Text, cases, and materials* (7th ed.). Oxford University Press.

Craig, P., & De Búrca, G. (Eds.). (2021). *The evolution of EU law* (3rd ed.). Oxford University Press.

d'Alfonso, A. (2020). *InvestEU programme: The EU's new investment support scheme*. Policy Commons. https://policycommons.net/artifacts/1426873/investeu-programme/2041448/

de Oliveira Neves, R. (2022). The EU taxonomy regulation and its implications for companies. In P. Câmara & F. Morais (Eds.), *The Palgrave handbook of ESG and corporate governance* (pp. 249–265). Springer.

De Visscher, C., Maiscocq, O., & Varone, F. (2008). The Lamfalussy reform in the EU securities markets: Fiduciary relationships, policy effectiveness and balance of power. *Journal of Public Policy, 28*(1), 19–47.

Dedman, M. (2010). *The origins and development of the European Union 1945-2008—A history of European integration* (2nd ed.). Routledge.

Delreux, T. (2018). Multilateral environmental agreements: A key instrument of global environmental governance. In C. Adelle, K. Biedenkopf, & D. Torney (Eds.), *European Union external environmental policy* (pp. 19–38). Springer International Publishing. https://doi.org/10.1007/978-3-319-60931-7_2

Deng, Y. (2007). The effects of patent regime changes: A case study of the European patent office. *International Journal of Industrial Organization, 25*(1), 121–138.

Duff, A., Pinder, J., & Pryce, R. (Eds.). (2002). *Maastricht and beyond*. Routledge. https://doi.org/10.4324/9780203427125

Duke, S. (2008). The Lisbon treaty and external relations. *Eipascope, 2008*(1), 1–6.

Dür, A. (2008). Bargaining power and trade liberalization: European external trade policies in the 1960s. *European Journal of International Relations, 14*(4), 645–669. https://doi.org/10.1177/1354066108097556

Dyer, H. C. (2017). Green theory. In S. McGlinchey, R. Walters, & C. Scheinpflug (Eds.), *International relations theory* (Vol. 2, pp. 84–90). E-International Relations. https://www.e-ir.info/publication/international-relations-theory/

EIT Regional Innovation Scheme. (2019, August 12). European Institute of Innovation & Technology (EIT). https://eit.europa.eu/our-activities/eit-regional-innovation-scheme

Elzen, B., Geels, F. W., & Green, K. (Eds.). (2004). *System innovation and the transition to sustainability: Theory, evidence and policy*. Edward Elgar.

European Commission. (2020). *Special Eurobarometer 501: Attitudes of European citizens towards the environment* (Special Eurobarometer 501). European Commission. https://data.europa.eu/data/datasets/s2257_92_4_501_eng?locale=en

European Union. (2020). *Shareholder rights directive*. https://eur-lex.europa.eu/legal-content/EN/TXT/HTML/?uri=LEGISSUM:l33285

Fabrizi, A., Guarini, G., & Meliciani, V. (2018). Green patents, regulatory policies and research network policies. *Research Policy*, *47*(6), 1018–1031. https://doi.org/10.1016/j.respol.2018.03.005

Fairhurst, J. (2016). *Law of the European Union* (11th ed.). Pearson Education.

Fernández, Y. F., López, M. F., & Blanco, B. O. (2018). Innovation for sustainability: The impact of R&D spending on CO_2 emissions. *Journal of Cleaner Production, 172*, 3459–3467.

Filipec, O. (2017). (In)efficiency of EU common foreign and security policy: Ukraine, Brexit, Trump and beyond. *Slovak Journal of Political Sciences, 17*(3–4). http://sjps.fsvucm.sk/index.php/sjps/article/view/7

Finotelli, C. (2018). Southern Europe: Twenty-five years of immigration control on the waterfront. In A. R. Servent & F. Trauner (Eds.), *The Routledge handbook of justice and home affairs research* (1st ed., pp. 240–252). Routledge. https://doi.org/10.4324/9781315645629-20

Franco, C., & Marin, G. (2017). The effect of within-sector, upstream and downstream environmental taxes on innovation and productivity. *Environmental and Resource Economics, 66*(2), 261–291.

Geels, F. W. (2004). From sectoral systems of innovation to socio-technical systems. *Research Policy, 33*(6–7), 897–920. https://doi.org/10.1016/j.respol.2004.01.015

Goddard, J., Liu, H., Molyneux, P., & Wilson, J. O. (2013). Do bank profits converge? *European Financial Management, 19*(2), 345–365.

Gouardères, F. (2021). *Innovation Policy* (Fact Sheets on the European Union) [Fact Sheets on the European Union]. European Parliament. https://www.europarl.europa.eu/factsheets/en/sheet/67/innovation-policy

Grahl, J. (2009). *Global finance and social Europe*. Edward Elgar Publishing.

Haastrup, T., Mah, L., & Duggan, N. (Eds.). (2021). *The Routledge handbook of EU-Africa relations*. Routledge.

Hedberg, A., & Šipka, S. (2022). The role of European Union policies in accelerating the green transition. *Field Actions Science Reports, Special Issue 24*, 86–91.

Hinchliffe, D., & Akkerman, F. (2017). Assessing the review process of EU Ecodesign regulations. *Journal of Cleaner Production, 168*, 1603–1613. https://doi.org/10.1016/j.jclepro.2017.03.091

Hooghe, L., & Marks, G. (2009). A postfunctionalist theory of European integration: From permissive consensus to constraining dissensus. *British Journal of Political Science, 39*(1), 1–23.

Horbach, J. (2008). Determinants of environmental innovation—New evidence from German panel data sources. *Research Policy, 37*(1), 163–173.

Hörber, T. (2006). *The foundations of Europe: European integration ideas in France*. VS Verlag für Sozialwissenschaften.

Hörber, T. (2012). *The origins of energy and environmental policy in Europe: The beginnings of a European environmental conscience*. Routledge.

Hörber, T., & Weber, G. (2022). *The European environmental conscience in EU politics: A developing ideology*. Routledge.

Howarth, D., & Quaglia, L. (2013). Banking on stability: The political economy of new capital requirements in the European Union. *Journal of European Integration, 35*(3), 333–346.

Iking, B. (2009). Benchmarking innovation performance on the regional level: Approach and policy implications of the European innovation scoreboard for countries and regions. In *Innovation, employment and growth policy issues in the EU and the US* (pp. 245–271). Springer.

Jänicke, M., & Wurzel, R. K. (2019). Leadership and lesson-drawing in the European Union's multilevel climate governance system. *Environmental Politics, 28*(1), 22–42.

Jiménez Sánchez, M., & Lafuente, R. (2010). Defining and measuring environmental consciousness. *Revista Internacional de Sociología, 68*(3), 731–755. https://doi.org/10.3989/ris.2008.11.03

Jones, E., Kelemen, R. D., & Meunier, S. (2016). Failing forward? The Euro crisis and the incomplete nature of European integration. *Comparative Political Studies, 49*(7), 1010–1034.

Kastrinos, N., & Weber, K. M. (2020). Sustainable development goals in the research and innovation policy of the European Union. *Technological Forecasting and Social Change, 157*, 120056. https://doi.org/10.1016/j.techfore.2020.120056

Kellerbauer, M., Klamert, M., & Tomkin, J. (2019). *The EU treaties and the charter of fundamental rights: A commentary.* Oxford University Press.

Kesidou, E., & Demirel, P. (2012). On the drivers of eco-innovations: Empirical evidence from the UK. *Research Policy, 41*(5), 862–870.

Kindleberger, C. P. (2015). *A financial history of Western Europe.* Routledge.

Knodt, M. (Ed.). (2003). *Understanding the European Union's external relations* (1. publ). Routledge.

Koehler, K. (2010). European foreign policy after Lisbon: Strengthening the EU as an international actor. *Caucasian Review of International Affairs, 4*(1), 57–72.

Kollmuss, A., & Agyeman, J. (2002). Mind the gap: Why do people act environmentally and what are the barriers to pro-environmental behavior? *Environmental Education Research, 8*(3), 239–260. https://doi.org/10.1080/13504620220145401

Koolen, C. (2019). Vacuum cleaner energy labels and misleading commercial practices: EU consumers left in the dust? *Journal of European Consumer and Market Law, 8*(2), 82–88. https://doi.org/EuCML2019014

Korres, G. M. (2013). The European national and regional systems of innovation. In *The innovation union in Europe.* Edward Elgar Publishing.

Krug, M., Di Nucci, M. R., Caldera, M., & De Luca, E. (2022). Mainstreaming community energy: Is the renewable energy directive a driver for renewable energy communities in Germany and Italy? *Sustainability, 14*(12), 7181. https://doi.org/10.3390/su14127181

Kuenzer, J. (2021). A growing European environmental conscience. In T. C. Hörber & G. Weber (Eds.), *The European environmental conscience in EU politics* (1st ed., pp. 185–206). Routledge, Taylor & Francis Group.

Kuhn, T. (2019). Grand theories of European integration revisited: Does identity politics shape the course of European integration? *Journal of European Public Policy, 26*(8), 1213–1230.

Laffan, B., & Mazey, S. (2006). European integration: The European Union—Reaching an equilibrium? In J. Richardson, *European Union: Power and policy-making* (pp. 30–53). Routledge.

Latoszek, E. (2021). Fostering sustainable development through the European digital single market. *Economics and Business Review, 7*(1), 68–89. https://doi.org/10.18559/ebr.2021.1.5

Leuffen, D., Rittberger, B., & Schimmelfennig, F. (2022). *Integration and differentiation in the European Union: Theory and policies.* Springer.

Mahlberg, B., Luptacik, M., & Sahoo, B. K. (2011). Examining the drivers of total factor productivity change with an illustrative example of 14 EU countries. *Ecological Economics, 72*, 60–69.

McNeill, J. (2021). Exporting environmental objectives or erecting trade barriers in recent EU free trade agreements. *Australian and New Zealand Journal of European Studies, 12*(1). https://doi.org/10.30722/anzjes.vol12.iss1.15077

Meiser, J. W. (2022). Liberalism. In S. McGlinchey, R. Walters, & D. Gold (Eds.), *International relations theory* (pp. 22–27). E-International Relations. https://www.e-ir.info/publication/international-relations-theory/

Monar, J. (2004). The EU as an international actor in the domain of justice and home affairs. *European Foreign Affairs Review, 9*, 195–415.

Murray, P. (2016). EU–Australia relations: A strategic partnership in all but name? *Cambridge Review of International Affairs, 29*(1), 171–191. https://doi.org/10.1080/09 557571.2015.1015487

Oberthür, S., & Roche Kelly, C. (2008). EU leadership in international climate policy: Achievements and challenges. *The International Spectator, 43*(3), 35–50.

OECD. (2011). *Promoting technological innovation to address climate change.* https:// www.oecd.org/env/consumption-innovation/49076220.pdf

Parenti, R. (2021). Financial services policy | Fact Sheets on the European Union. *European Parliament.* https://www.europarl.europa.eu/factsheets/en/sheet/83/financial-services-policy

Pegkas, P., Staikouras, C., & Tsamadias, C. (2019). Does research and development expenditure impact innovation? Evidence from the European Union countries. *Journal of Policy Modeling, 41*(5), 1005–1025.

Pittaluga, G. B. (2016). The European monetary system. In D. Preda (Ed.), *The history of the European monetary union—Comparing strategies amidst prospects for integration and national resistance* (pp. 89–106). P.I.E. Peter Lang.

Pollex, J., & Lenschow, A. (2018). Surrendering to growth? The European Union's goals for research and technology in the Horizon 2020 framework. *Journal of Cleaner Production, 197*, 1863–1871. https://doi.org/10.1016/j.jclepro.2016.10.195

Porter, M. E., & Van der Linde, C. (1995). Toward a new conception of the environment-competitiveness relationship. *Journal of Economic Perspectives, 9*(4), 97–118.

Quaglia, L. (2010). *Governing financial services in the European Union: Banking, securities and post-trading.* Routledge.

Quaglia, L. (2013). Financial regulation and supervision in the European Union after the crisis. *Journal of Economic Policy Reform, 16*(1), 17–30.

Razzaque, J. (2012). *Environmental governance in Europe and Asia: A comparative study of institutional and legislative frameworks.* Routledge.

Rexhäuser, S., & Rammer, C. (2014). Environmental innovations and firm profitability: Unmasking the Porter hypothesis. *Environmental and Resource Economics, 57*(1), 145–167.

Ricciardiello, L., Leja, M., & Ollivier, M. (2021). Horizon Europe, the new programme for research & innovation: Which opportunities for GI research in the years to come? *United European Gastroenterology Journal, 9*(3), 407–409. https://doi.org/10.1002/ ueg2.12073

Richardson, J. (2006). Policy-making in the EU: interests, ideas and garbage cans of primeval soup. In J. Richardson (Ed.), *European Union: Power and policy-making* (pp. 2–29). Routledge.

Rootes, C. (2002). Global visions: Global civil society and the lessons of European Environmentalism. *Voluntas: International Journal of Voluntary and Nonprofit Organizations, 13*(4), 411–429.

Rootes, C. (2003). *Environmental protest in western Europe.* Oxford University Press on Demand.

Rubashkina, Y., Galeotti, M., & Verdolini, E. (2015). Environmental regulation and competitiveness: Empirical evidence on the Porter Hypothesis from European manufacturing sectors. *Energy Policy, 83*, 288–300.

Sarti, M., & St. John, S. K. (2019). Raising long-term awareness: EU environmental policy and education. In S. K. St. John & M. Murphy (Eds.), *Education and public policy in the European Union* (pp. 165–181). Springer International Publishing. https://doi.org/10.1007/978-3-030-04230-1_8

Schäfer, A. (2006). Beyond the community method: Why the open method of coordination was introduced to EU policy-making. In R. Holzhacker & M. Haverland (Eds.), *European research reloaded: Cooperation and europeanized states integration among europeanized states* (pp. 179–202). Springer Netherlands. https://doi.org/10.1007/1-4020-4430-5_8

Schaub, A. (2005). The Lamfalussy process four years on. *Journal of Financial Regulation and Compliance, 13*(2), 110–120.

Schenk, C. R. (2010). The regulation of international financial markets from the 1950s to the 1990s. In J. Reis & S. Battilossi (Eds.), *State and financial systems in Europe and the USA* (pp. 149–166). Routledge.

Schleich, J., Durand, A., & Brugger, H. (2021). How effective are EU minimum energy performance standards and energy labels for cold appliances? *Energy Policy, 149*, 112069.

Schoenmaker, D. (2016). The banking union: An overview and open issues. In T. Beck & B. Casu (Eds.), *The Palgrave handbook of european banking* (pp. 451–474). Palgrave Macmillan.

Schulze, K., & Tosun, J. (2013). External dimensions of European environmental policy: An analysis of environmental treaty ratification by third states. *European Journal of Political Research, 52*(5), 581–607.

Scruggs, L., & Benegal, S. (2012). Declining public concern about climate change: Can we blame the great recession? *Global Environmental Change, 22*(2), 505–515. https://doi.org/10.1016/j.gloenvcha.2012.01.002

Sfârlog, T.-V. (2020). Critical appraisal of the recent case-law of the European Union intellectual property office (EUIPO). *International Conference Knowledge-Based Organization, 26*(2), 228–231. https://doi.org/10.2478/kbo-2020-0081

Shahbaz, M., Nasir, M. A., & Roubaud, D. (2018). Environmental degradation in France: The effects of FDI, financial development, and energy innovations. *Energy Economics, 74*, 843–857.

Sharma, K., & Bansal, M. (2013). Environmental consciousness, its antecedents and behavioural outcomes. *Journal of Indian Business Research, 5*(3), 198–214. https://doi.org/10.1108/JIBR-10-2012-0080

Shyrokykh, K. (2022). Help your neighbor, help yourself: The drivers of European Union's climate cooperation in trans-governmental networks with its neighbors. *Governance, 35*(4), 1095–1118. https://doi.org/10.1111/gove.12646

Silvestre, B. S., & Țîrcă, D. M. (2019). Innovations for sustainable development: Moving toward a sustainable future. *Journal of Cleaner Production, 208*, 325–332.

Smith, A., Voß, J.-P., & Grin, J. (2010). Innovation studies and sustainability transitions: The allure of the multi-level perspective and its challenges. *Research Policy, 39*(4), 435–448. https://doi.org/10.1016/j.respol.2010.01.023

Sørensen, G., Møller, J., & Jackson, R. (2022). *Introduction to international relations: Theories and approaches*. Oxford University Press.

Talens Peiró, L., Polverini, D., Ardente, F., & Mathieux, F. (2020). Advances towards circular economy policies in the EU: The new ecodesign regulation of enterprise servers. *Resources, Conservation and Recycling, 154*, 104426. https://doi.org/10.1016/j.resconrec.2019.104426

Tallon, A. (2020). *Head and heart: Affection, cognition, volition, as truine consciousness.* Fordham University Press. https://doi.org/10.1515/9780823295791

Tendero, M. (2021). The time has come! The development of the European environmental conscience. Evidence from the Eurobarometer surveys from 1974 to 2020. In T. C. Hörber & G. Weber (Eds.), *The European environmental conscience in EU politics* (1st ed., pp. 185–206). Routledge, Taylor & Francis Group.

Thatcher, M. (2006). European regulation. In J. Richardson (Ed.), *European Union: Power and Policy-Making*. Routledge.

Theys, S. (2017). Constructivism. In S. McGlinchey, R. Walters, & C. Scheinpflug (Eds.), *International relations theory* (Vol. 2, pp. 36–41). E-International Relations. https://www.e-ir.info/publication/international-relations-theory/

Trærup, S. (2022). The role of climate technologies in green transition pathways. *Field Actions Science Reports, Special Issue 24*, 28–31.

Triguero, A., Moreno-Mondéjar, L., & Davia, M. A. (2013). Drivers of different types of eco-innovation in European SMEs. *Ecological Economics, 92*, 25–33.

Ullrich, H. (2002). Patent protection in Europe: Integrating Europe into the community or the community into Europe? *European Law Journal, 8*(4), 433–491. https://doi.org/10.1111/1468-0386.00161

UNFCCC. (2015). *Adoption of the Paris Agreement.* https://unfccc.int/sites/default/files/english_paris_agreement.pdf

Vogler, J. (2005). The European contribution to global environmental governance. *International Affairs, 81*(4), 835–850.

Weber, C. S. (2015). The Euro crisis: Causes and symptoms. *Estudios Fronterizos, 16*(32), 150–172.

Weber, M., Lamprecht, K., & Biegelbauer, P. (2019). *The Shaping a new understanding of the impact of horizon Europe: The roles of the European Commission and member states.* fteval – Platform for Research and Technology Policy Evaluation. https://doi.org/10.22163/fteval.2019.347

Wouters, J., Bruyninckx, H., Basu, S., & Schunz, S. (2012). *The European Union and multilateral governance: Assessing EU participation in united nations human rights and environmental fora.* Springer.

Part I

The Influence of Sustainability on the External Relations of the European Union

1 Financial Enforcement Instruments of Sustainability

Frauke Pipart

Introduction

In international environmental negotiations, the European Union (EU) is often praised as leading-by-example or being a leader, pushing for ambitious goals to protect the environment (Bäckstrand & Elgström, 2013; Delreux & Pipart, 2021; Vogler, 2005). The EU's leadership ambition is supported by its growing expertise in environmental policies, which has evolved since the 1970s, supporting the presence of the European Environmental Conscience (Hoerber et al., 2022). The EU's strong desire to use multilateralism to solve environmental issues is explained by the "high level of environmental protection in the EU and the fact that European companies are subject to strong international competition" (Delreux & Happaerts, 2016, p. 244). This increases the EU's wish to implement the same level of environmental protection globally (Kelemen, 2010). Nevertheless, in particular, in the context of climate change negotiations, scholars have identified varying levels of EU ambition. For example, the EU downscaled its ambition in the lead-up to the Paris climate conference in 2015 to put forward a more realistic position (Oberthür & Groen, 2018). Moreover, Groen (2018) identifies varying ambitions in the context of the Convention on Biological Diversity (CBD), as the EU defended moderately ambitious positions on the negotiations towards the Cartagena Protocol, but rather unambitious positions with regard to the Nagoya Protocol. Biedenkopf et al. (2018) argue that even though they observe a tendency of the EU to have a rather high ambition on environmental policies, this does not mean that the EU's ambition can be described as high in all instances. By analysing the EU's ambition, this chapter contributes to the literature on the EU's leadership. The EU's performance as a normative environmental leader depends on its ambition and can be put in question when the EU's ambition decreases (Burns & Tobin, 2018). Hence, studying to what extent the EU externalises its internal environmental ambition is a crucial factor, contributing to the understanding of the European environmental conscience, which forms the key methodology of this book. As more and more energy and environmental policies have been introduced at the EU level, the EU has become an international expert in the field and usually tries to transfer these internal policies to

DOI: 10.4324/9781032656359-3

the global level. This chapter wishes to address two shortcomings. First, most studies directly link the EU's ambition with its international leadership and goal achievement. Bäckstrand and Elgström (2013) argue that the EU had difficulties in promoting its ambitious positions because it was largely isolated and tried to make too normative arguments for a highly ambitious agreement. When the EU favoured less ambitious policies one year later at the Cancun conference, it managed to achieve many more of its goals, given that these positions could more easily find approval (Groen et al., 2012). Hence, it seems that the EU should rather favour less-ambitious policies to be successful. Yet, the EU does want to drive environmental action forward and hence often chooses to still push for high-ambitious policies. Currently, studies do not explain why and to what extent the EU's ambition varies. In addition, Liefferink and Wurzel (2017) argue that the literature has so far attributed little importance to changes over time in the EU's ambition. Hence, the literature currently misses a proper understanding of how and why variation in the EU's ambition is observed, a black box this chapter wishes to unpack.

Second, most literature on international environmental negotiations focuses on the EU's ambition vis-à-vis climate change. Scholars examine the evolution of the EU from the Kyoto Protocol until the Paris Agreement and find that the EU regularly adapted its ambition to changing external circumstances (Bäckstrand & Elgström, 2013; Parker et al., 2017). These changes include changes in the positions of the United States and "a growing rift between BASIC countries and low-lying islands" (Bäckstrand & Elgström, 2013, p.1381). In these negotiations, the EU usually defends ambitious policies, although this ambition varied over time. However, environmental concerns including biodiversity loss or chemicals and waste management remain largely understudied. By assessing the EU's ambition in the CBD, as well as the Basel (BC), Rotterdam (RC), and Stockholm (SC) Conventions, this chapter broadens the view on environmental negotiations beyond climate change. It can therefore test if the observed variation in the EU's ambition on climate is equally observed in other environmental conventions.

The chapter answers the following research questions: (1) How does the EU's ambition vary across international environmental conventions? (2) How did the EU's level of ambition in these conventions evolve over time? To compare the EU's ambition, this chapter concentrates on the negotiations between 2012 and 2019. Research is based on the analysis of information notes, which are published by the EU, and which include the statements given by the EU and its member states at the Conferences of the Parties (COPs). The results show that the EU's ambition is usually positive, that is, the EU defends weak, moderate, or highly ambitious positions. Its ambition is highest in the Rotterdam Convention and lowest in the Basel Convention. Contrary to previous findings from the climate literature, the EU's ambition only evolved to a very limited extent in recent years. These findings support the presence of a European environmental conscience, with the EU's strong desire to develop innovative, ambitious solutions and support progress at the global level.

The remainder of this chapter is structured as follows: the second section reviews the existing literature on the EU in international negotiations, and the third section outlines the theoretical framework and methodology. The fourth section explains the assessment framework for EU ambition. Then, the fifth section outlines the data,the sixth section concentrates on the assessment of the EU's ambition, and the seventh section discusses potential explanations for its ambition. The last section concludes the chapter.

EU Leadership and International Environmental Negotiations

The EU has a preferential status in many multilateral conventions, as it often manages to obtain the same status as countries usually have. This development started with the entry into force of the Lisbon Treaty in 2009. The EU took on legal personality and could thus upgrade its status from being an observer to being a party to several multilateral environmental agreements (MEAs) (Delreux, 2013; Laatikainen, 2010). This includes the Basel, Rotterdam, and Stockholm Conventions, as well as the CBD. The EU and all its member states are parties to these conventions. In all four conventions, the European Commission ensures that a common position is established ahead of the negotiations, and – in many cases – the Commission is negotiating on behalf of the EU and its member states. Hence, the European Commission has indeed developed a strong expertise in environmental policies and is perceived to be a legitimate actor in the negotiations. This supports the exist of a European environmental conscience at the global level (Hoerber & Weber, 2022).

Delreux and Happaerts (2016) argue that the EU has developed the most ambitious internal framework worldwide to regulate environmental issues and that the EU desires to expand this level of environmental protection to other countries in the world. The European Commission has tried to successfully present the EU as a leader in international climate negotiations (Wurzel et al., 2017). The observation that the EU is an ambitious environmental actor does however neither automatically mean that its policies are sufficiently ambitious to address the environmental problems effectively nor that its position allows it to achieve the goals of the respective international agreement (Biedenkopf et al., 2018).

At the international level, the EU favours highly ambitious policies on climate change since the early 1990s and the negotiations of the Kyoto Protocol (Bäckstrand & Elgström, 2013). Yet, the EU has moderated its ambition after the failure at the Copenhagen Conference, 2009, to adapt to the changing international context (Liefferink & Wurzel, 2017; Oberthür & Groen, 2015). In the successful 2015 negotiations in Paris, the EU's position was moderately ambitious and thereby "much closer to those of the US, China and other major emitters" than in the Copenhagen COP (Oberthür & Groen, 2018, p. 719). Since then, the EU has "urged parties to upgrade their mitigation ambitions" (Oberthür & Dupont, 2021, p. 1101). Hence, the EU has defended ambitious, yet realistic positions since 2010, in particular when compared to other major players in climate change negotiations.

Groen (2018) has previously identified the EU's level of ambition in the initial negotiations of the CBD and its protocols. She claims that whereas the EU had moderately ambitious positions in the negotiation on the CBD and the Cartagena protocol, it was rather unambitious in the negotiations on the Nagoya protocol. However, she does not provide an explanation for these differences. Analysing global chemicals governance, including the Basel, Rotterdam, and Stockholm (BRS) Conventions, Biedenkopf (2016) finds that the EU often favours more ambitious positions than other parties. Discussing the implementation of the RC, Jansen and Dubois (2014) argue that the EU, contrary to many developing countries, largely uses the principles of the conventions in its decision-making. Assessing the EU's ambition in the most recent negotiations of the BRS and CBD, Delreux and Pipart (2021) find that the EU is overall perceived to have a very high level of environmental ambition by delegates of these negotiations. Yet, studying at the issue level, Pipart (2022) identifies both high and low levels of ambition in these negotiations.

A growing literature analyses the EU's climate ambition at the EU or even member states level. In doing so, scholars frequently study the ambition of legislation adopted by the EU. Following a range of crises, including the global financial crisis 2007–2008, the EU's ambition on climate issues waned to some extent, even though more policies were created (Burns & Tobin, 2020). This mirrors previous studies arguing for a crisis impact on the EU's environmental ambition more generally (Burns & Tobin, 2018). Former climate pioneers were reluctant to push for highly ambitious policies, making it more difficult for the EU to keep up its ambition (Burns et al., 2020; Wurzel et al., 2017). However, Gravey and Moore (2018) argue that in most cases, the EU policy intensity, that is, the number of policies adopted, "increased both before and after 2008". It remains unclear if the crisis effect equally spread to the EU's ambition at the international level.

Following the literature review, I expect the EU to show a highly ambitious position in all four conventions, with a relatively low variation between them. As all conventions focus on environmental issues, I expect that the EU defends similar levels of ambition on most issues. Given that the selected timeframe starts after the economic crisis, I expect that variation will either be low, or the EU's ambition will have slightly increased over time.

Theoretical Framework and Methodology

In this chapter, the EU is understood as both the EU institutions and the EU member states. Given that environmental matters are shared competences, both the EU and its member states are competent in the field. EU institutions, mostly officials from the European Commission, represent the EU in matters related to Union competences (Keukeleire & Delreux, 2014). If an issue touches rather the competences of the member states, it is often the rotating Presidency who speaks for the EU. Yet, a common procedure is that the European Commission presents "recommendations to the Council, which then decides on who will represent the Union" at the international level (Corthaut & Van Eeckhoutte, 2012, p. 151).

This chapter primarily contributes to the understanding of the European environmental conscience at the international level. The European Commission is of particular relevance, as it "introduced both energy and environmental policies into the EU policy canon and hence built an institutional framework, which reflected EU expertise in these fields" (Hoerber et al., 2022, p. 77). In international negotiations, the Commission, together with the EU member states tries to transfer these internal policies to the global level. Therefore, this chapter studies to what extent the EU proposes innovative solutions and commitments to progress in international environmental negotiations.

Assessment Framework for EU Ambition

To operationalise the existence of a European environmental conscience in international environmental negotiations, I study the EU's ambition. An actor is considered ambitious if he or she wants to achieve more stringent policies than the current status-quo provides (Burns et al., 2020). Ambition is then defined as "the extent to which a country or regional group prefers a negotiation outcome that will lead to a high level of environmental protection" (Ohler & Delreux, 2021, p. 11). To assess the EU's level of ambition, scholars propose to assess the EU's ambition in international negotiations in relative terms, that is, in comparison to other actors. This chapter adopts a different approach, by systematically comparing the EU's ambition on different issues under negotiation. Hence, this level of ambition is not in relation to other actors, but only a comparison of the EU with itself. This allows for a better comparison over time, as changing preferences among the EU's negotiation partners will not affect the results.

Burns et al. (2020) have introduced a typology of the EU's ambition in order to assess proposals for EU environmental legislation. Their typology is used as a basis for my assessment framework but needs to be adapted, as the definitions do not fit the context of positions presented at international negotiations (see Table 1.1). As in the typology, I create three categories for "positive" ambition, one for "negative" ambition and one neutral category. This assumes that it is relatively unlikely that the EU will at all produce positions with a negative ambition, given its usual leadership aspirations on environmental topics. My assessment framework closely follows the typology of Burns et al. (2020) but the definitions are slightly changed. These adaptions are highlighted in italics in Table 1.1.

Data

To assess the EU's ambition, I have selected four environmental conventions: CBD, BC, RC, and SC. These four conventions were chosen to counter the strong climate bias in the literature. The EU's positions on the loss of biodiversity as well as the production and use of chemicals and wastes have been far less studied. Focusing on these environmental conventions allows to test if similar

Table 1.1 Assessment framework for EU ambition

	Level of ambition
5	High ambition: *goal is to adopt* ambitious, binding targets and clear, ambitious deadlines.
4	Moderate ambition: *goal is to adopt* targets that are more ambitious than the status-quo, but less than the high ambition. Deadlines *are either not included* or give very long timeframes.
3	Weak ambition: *goal is to increase the ambition of the status-quo, but changes remain marginal and have no binding commitments. Mainly rhetorical commitments and changes that do not have a strong practical impact.*
2	Neutral ambition: *goal is to* maintain the status-quo. *Requested changes are* typically editorial or neutral amendments.
1	Negative ambition: *goal is to* weaken the status-quo *of the convention, by requesting changes that would lead to lower environmental protection or delay them* (for example, by extending deadlines).

Source: Adapted from Burns et al. (2020).

dynamics than in climate negotiations are observed or if the EU's ambition does significantly differ. I study these conventions between 2012 and 2019. The seven-year period was chosen for two reasons. First, focusing on the years 2012 to 2019 allows to keep many external factors constant. The EU's external representation in these negotiations is constant since the adoption of the Lisbon Treaty in 2009 and the EU is a party to all four conventions. In addition, the BC, RC, and SC Conventions are negotiated at one single meeting since the ordinary COPs in 2013, that is, the first COP in the sample. Second, the choice for the time period has been made for practical reasons, as it is only since that 2012 that the EU has consistently published the plenary statements of the COPs. The study ends in 2019, thereby studying COPs that happened prior to the COVID-19 pandemic. In all conventions, the main negotiations take place at the COP every two years. Thus, the sample includes four COPs per convention. The BC, RC, and SC COPs are conducted simultaneously at one meeting, yet separate EU positions are established. The agenda items that are negotiated together under the so-called synergies process are not part of the analysis. Even though the conventions negotiate at a single meeting, they nonetheless have separate agendas and different parties are involved, as membership differs. Hence, it is worthwhile to study the EU's ambition separately for the three conventions. I do not expect to witness an impact of the parallel meeting on the EU's ambition.

The EU's ambition is assessed based on the position of the EU at the start of the negotiations. These positions are outlined in the information notes published by the General Secretariat of the Council of the EU. The documents outline every plenary statement on all issues, delivered by the EU in the meetings. The notes are published after the conclusion of the COPs. By coding the information notes, the chapter is able to compare the EU's ambition both among conventions and over time. The EU's ambition is assessed at issue-specific level. An issue is a specific topic negotiated by the COP and is identified

Table 1.2 List of interviews

1.	Interview BRS	20/05/2019	Skype
2.	Interview BRS	21/05/2019	Skype
3.	Interview BRS	11/06/2019	Skype
4.	Interview BRS	14/06/2019	Skype
5.	Interview BC	19/06/2019	Skype
6.	Interview CBD	10/07/2019	Skype
7.	Interview CBD	09/08/2019	Skype
8.	Interview CBD	13/01/2021	Skype
9.	Interview SC	18/01/2021	MS Teams
10.	Interview BC	12/02/2021	MS Teams
11.	Interview SC	17/02/2021	Webex
12.	Interview BC	01/03/2021	MS Teams
13.	Interview RC	04/03/2021	MS Teams
14.	Interview RC	09/03/2021	MS Teams
15.	Interview CBD	10/03/2021	MS Teams
16.	Interview CBD	10/03/2021	MS Teams
17.	Interview BC	11/03/2021	MS Teams
18.	Interview BC	25/03/2021	MS Teams
19.	Interview CBD	20/04/2021	MS Teams

Source: Own visualisation

based on the EU documents. They are structured by headlines separating each issue. After attributing a level of ambition per issue, this chapter proceeds to calculate frequencies, in order to determine the level of ambition per COP. The analysis of the EU's ambition will proceed in two steps. First, I compare the level of ambition of the EU between the four conventions. Second, I analyse the EU's ambition over time per convention.

To provide some tentative explanations of the EU's ambition, this chapter pursues an inductive approach. I relied on interviews with EU delegates who participated in recent COPs, as well as reporters from the Earth Negotiation Bulletin (ENB). 19 interviews allow to outline potential reasons for difference in the EU's ambitions, both between the conventions and over time. The interviews were conducted virtually in 2019 and 2021 (Table 1.2).

Findings

Differences between Conventions

A first observation when comparing the conventions is that the EU did not defend a negative ambition in any COP (see Figure 1.1). This is unsurprising, as a negative ambition would mean that the EU actively tries to weaken status-quo of the convention. Moreover, even a neutral EU ambition is rarely observed. This mirrors previous studies arguing that the EU usually defends rather ambitious positions in the conventions under study (Biedenkopf, 2018; Delreux & Pipart, 2021; Groen, 2018). Yet, in the Basel Convention, the EU defends a neutral ambition in 20% of the issues. For example, on the technical

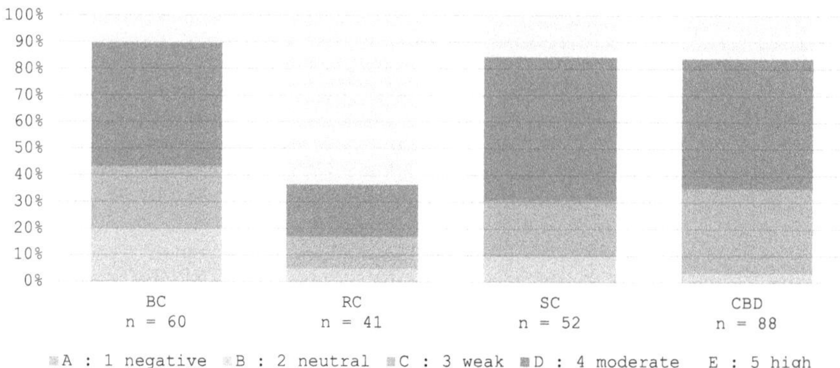

Figure 1.1 Comparison between conventions.

guidelines on incineration, the EU argued that it was "hesitant as regards the need for updating of these guidelines" (General Secretariat of the Council, 2017, p. 15). In the Stockholm Convention, 10% of the cases show a neutral EU ambition and in the Rotterdam Convention and CBD this is even less common.

The Rotterdam Convention stands out by showing the highest levels of ambition of the EU. In 63% of all issues, the EU defended a highly ambitious position. For instance, the EU strongly supported the listing of chrysotile asbestos in all four COPs under study (see for example General Secretariat of the Council, 2017). The other three Conventions show a more similar pattern. On all issues, the EU most often defends a moderate ambition. In these cases, the EU typically supports the draft decision and proposes minor amendments to it. I will provide some tentative explanations for these differences in the discussion section.

Differences Over Time

To compare differences over time, Figure 1.2 offers an overview of the EU's ambition of each convention per COP. The ambition of the EU in all conventions shows some variation over time, yet to varying extents. Put differently, no cross-convention patterns emerge. In the Basel Convention, the EU's ambition was relatively similar in 2013, 2017, and 2019. In 2015 however, the EU's ambition was considerably more positive. The share of high and moderate ambition was higher compared to the other COPs.

The Rotterdam Convention shows a slight increase in the EU's ambition over the years. At all four COPs, the EU defended a high ambition in at least 50% of the cases. Moderately more variation is observed for the other categories, but the general tendency is that the EU usually defends moderate or high ambitions in the RC.

In the Stockholm Convention, the EU's ambition has been relatively consistent over the years. Most often, the EU did defend high or moderate ambitions.

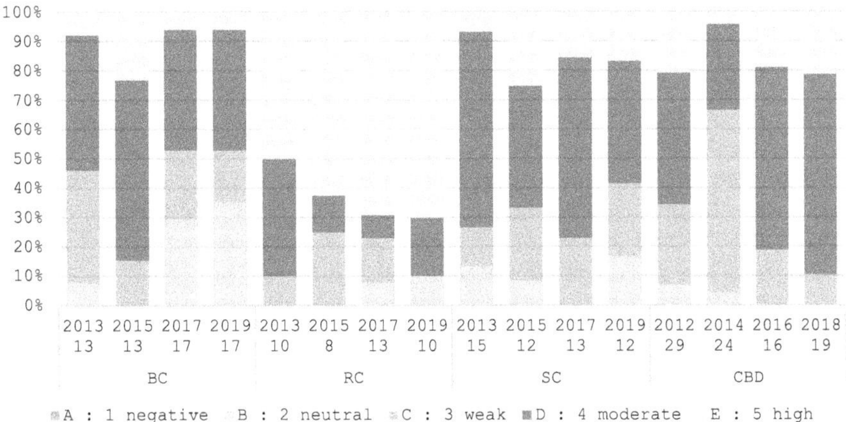

Figure 1.2 Comparison over time.

In 2015, the share of high-ambitious positions was highest, yet the EU then defended less often a moderate ambition compared to the years 2013 and 2017. Only in 2017, did the EU not defend a neutral ambition on any issue.

In the CBD, the EU's ambition was considerably lower in 2014 than in the years before and after. Exclusively in one case, the EU supported a highly ambitious position and most of the time, it favoured moderate positions. The years 2016 and 2018 look particularly similar, as the EU did not defend any neutral ambition and the share of weak, moderate, and high ambition is nearly the same.

Overall, there is no clear pattern in the EU's ambition over time. The EU's ambition varied between the years, and it is not possible to draw overall conclusions or provide predictions on the evolution of the EU's ambition in the upcoming years. However, changes are rather minor, and the EU's ambition is in general relatively constant.

Discussion

The following sections provide some tentative explanations for the EU's ambition, based on interviews with COP participants. These explanations are neither complete nor do they hold true for all changes in the EU's ambition. Yet, they can give some first hints on why the EU's ambition varies in international negotiations.

Internal EU policies

The EU's ambition frequently depends and evolves based on existing EU internal policies. If a regulation exists at EU level on a topic that is negotiated at a COP, the EU is pushing for adopting a similar – or less ambitious – position at the COP (Interview 1, 3). This argument is not novel. Delreux (2014, p. 1026) found that the EU's positions "at the international level often originate from the legislation existing on a particular environmental subdomain at the EU level". This can push the EU to defend both high and low-ambitious positions, depending on how ambitious its internal regulation is (Interview 12). For example, the EU's ambition on PFOA was low, because the parallel EU internal process had brought forward the additional need for exemption to the listing of PFOA (Interview 9, 11). Hence, the EU's ambition internally is usually mirrored by the EU's ambition at the COP. It has previously been argued that the EU has developed the most ambitious internal framework in the world, wishing to regulate environmental issues (Delreux & Happaerts, 2016). Hence, it is likely that existing EU policies will mainly increase the EU's ambition at the international level. If no policy exists at the EU level on an issue negotiated at the COP, the EU will most likely defend a low-ambitious position.

Nature of the Issue

Interviewees argued that the nature of the issue can determine the EU's ambition (Interviews 3, 6, 7). There are three observations made regarding the nature of the issue, which has an impact on the EU's ambition: the political nature of the issue, the financial implications, and the legally binding character of the agreement. Importantly, these differences matter only when comparing the EU's ambition on different issues. When ambition is compared between different actors (for example, Ohler & Delreux, 2021), the nature of the issue would be irrelevant, given that it is the same for all actors.

First, if the issue is highly political, the EU is more likely to wish to demonstrate its willingness to adopt an agreement and thus its ambition will be high (Interview 18). Low political issues, such as adopting new technical guidelines or amending them, usually does not push the EU to present high ambitions. The EU regularly defends a neutral ambition in relation to technical guidelines, for example, on e-waste or incineration (General Secretariat of the Council, 2017).

Second, if the issue has financial implications for the EU, it is more likely that the EU will show a low level of ambition (Interview 8, 19). For instance, the EU's ambition might be comparatively low or neutral when the decision is about establishing a new working group. Given that such decisions imply additional meeting costs, the EU in some cases is against this if it is considered unnecessary or too early in the process. In other cases, with financial implications, the EU might employ delaying tactics (Interview 8). This confirms previous observations. Oberthür and Groen (2017, p. 3) argued that "the EU was determined to prevent quantified long-term legal obligations and strongly advocated widening the circle of contributing countries", when parties discussed climate finance under the Paris Agreement.

Third, the legally binding character of the issue or the convention as a whole is important. The differences between the RC on the one hand and the BC, SC, and CBD on the other hand, can partially be explained by the issues that are negotiated in the RC. In the RC, parties list chemicals, which are then subject to the prior informed consent (PIC) procedure. The EU is usually strongly in favour of listing the chemicals that are proposed (Interview 3). Contrary to other conventions, the PIC does not prevent parties from exporting or importing these chemicals. This makes it easier for the EU to push for highly ambitious positions. The Stockholm Convention however "seeks to eliminate the production and use of chemicals" and is a legally binding convention (Downie & Templeton, 2016, p. 411). In recent years, the SC has been treating chemicals that are still in use and the EU is more often pushing for additional exemption that it used to, in order to avoid too strict implications (Interview 3). Similarly, the Basel Conventions tries to "prohibit the import of hazardous wastes or other wastes for disposal" (Secretariat of the Basel Convention, 2018, p. 10). Hence, compared to the RC, the consequences of decisions adopted under the BC and SC seem to be more important to the EU. In particular, when decisions at the COP create legally binding effects in the EU's own legal order, the EU's ambition will be lower.

Failure or Success of previous Negotiation Rounds

When the EU defended a moderate or high ambition but failed to adopt an agreement due to the preferences of other actors, its ambition usually remained constant at following COPs. In these cases, the EU favoured adopting new agreements on the issue, but due to other parties, no decision could be adopted. Hence, the EU carried its position on to the next COP, without changes. The best example of this case is the listing of chrysotile asbestos in the Rotterdam Convention. The issue was discussed for several COPs, and the EU was always in favour of listing asbestos in the RC. Likewise, the EU's ambition remained constant regarding the issue of adopting a compliance mechanism in the RC. The EU kept its high ambition during all years of negotiations, until it was adopted in 2019 (Interview 13, 14).

In addition, the EU's ambition might be lower when long-lasting, complicated discussions finally come to an agreement. For example, on the negotiations on digital sequence information in CBD, the EU's ambition is low because it does not want to risk reopening the Nagoya Protocol, which could only be concluded after many years of discussion (Interview 19). Similarly, the EU had difficulties putting forward a highly ambitious position on the modifications of Ecologically or Biologically Significant Marine Areas (EBSAs), as an initial decision was established in 2010, after long and complicated negotiations (Interview 15, 16).

Conclusion

Assessing the EU's ambition in four international environmental conventions, this chapter shows that the EU mostly defends positions with moderate to high ambitions. This observation confirms previous studies highlighting the EU's ambitious positions (Biedenkopf, 2018; Delreux & Pipart, 2021). However, the analysis also shows that the EU regularly defends a weak or even neutral ambition, in particular in the Basel Convention. In the Rotterdam Convention, the EU showed the highest level of ambition, often arguing that listing new chemicals is essential. In the Stockholm Convention and the CBD, the EU defends remarkably similar levels of ambition, most often it favours high or moderate ambitions. When analysing the EU's development over time, I could not identify clear patterns. Whereas the EU's ambition revealed some variation over the years, it did not evolve in a clear direction. Tentative explanations for the EU's ambition broadly revolve around three themes: internal EU policies, the nature of the issue, and the failure or success of previous negotiation rounds.

The EU's overall ambition across these conventions displays a strong desire for multilateral solutions to environmental challenges. The EU generally tries to push for more ambition compared to what already exists at the global level. By externalising existing EU policies and pushing for more ambition internationally, the EU is committed to making progress on global solutions.

These conclusions have critical implications for the literature on the European Environmental Conscience and the EU's international leadership. First, the chapter shows that the EU's ambition in the studied international environmental conventions has been relatively stable in recent years. The EU is a clear expert on environmental and energy policies, having a very advanced set of policies in these fields at the EU level. Hence, it generally tries to externalise these policies to the global level. Although only four conventions were analysed, this leads to broader reflections. Scholars have previously argued that the EU had been impacted by the economic and migration crisis, leading to lower ambitions and more variation in particular in the climate regime (Burns & Tobin, 2018). Yet, at least at the international level, the EU presents similar positions on key environmental topics since 2012 and if a crisis effect did indeed occur, it has been resolved since.

Second, the chapter points to the relevance of understanding the EU's level of ambition when studying the EU's leadership and influence in international environmental negotiations. At all COPs, I identified issues with weak, moderate, and high levels of ambition and in some also a neutral EU ambition. Variations in the EU's ambition can help explain differences in its leadership and influence. Hence, scholars of international negotiations should focus on analysing EU influence and leadership at issues-specific level and not only for COPs overall. The analysis reveals that the EU's ambition is not constant for a single COP, but rather for specific issues under negotiation throughout the years.

Whereas the chapter provided an overall assessment of the EU's ambition in four environmental conventions, it could only put forward tentative explanations as to why some variation was observed. Scholars could therefore focus on trying to confirm these tentative explanations with additional research. To do so, research at COP and EU level needs to be combined. Given that the EU's ambition is established at the EU level prior to the COP, studying EU and member states negotiations might reveal deeper insights into the reasons behind the EU's ambitions.

In addition, by focusing on four rather low politicised conventions and a timeframe of seven years, these conclusions should mainly be interpreted for the cases under study. It is likely that the low political nature of these conventions makes it easier for the EU to be ambitious. Hence, similar conventions might display a relatively similar level of ambition. Yet, for example, the climate regime, with its strong politicised nature, might display different features. The timeframe covers, as explained, the pre-COVID era. Subsequent research might equally be interested in considering the effect of the current COVID-19 pandemic on the EU's ambition in international environmental negotiations. The global pandemic and the postponement of the CBD and BRS COPs, might have had an effect on the EU. Moreover, as suggested above, combining the analysis of the EU's ambition at issue-level with its leadership and influence could provide promising insights.

Bibliography

Bäckstrand, K., & Elgström, O. (2013). The EU's role in climate change negotiations: From leader to 'leadiator'. *Journal of European Public Policy*, *20*(10), 1369–1386.

Biedenkopf, K. (2016). The EU in global chemicals governance. In I. Peters (Ed.), *The European Union's foreign policy in comparative perspective. Beyond the "actorness and power" debate* (pp. 61–79). Routledge.

Biedenkopf, K. (2018). Chemicals: Pioneering Ambitions with External Effects. In C. Adelle, K. Biedenkopf, & D. Torney (Eds.), *European Union external environmental policy: Rules, regulation and governance beyond borders* (pp. 189–208). Palgrave macmillan.

Biedenkopf, K., Torney, D., & Adelle, C. (2018). Conclusions. In C. Adelle, K. Biedenkopf, & D. Torney (Eds.), *European Union external environmental policy: Rules, regulation and governance beyond borders*. Palgrave macmillan.

Burns, C., Eckersley, P., & Tobin, P. (2020). EU environmental policy in times of crisis. *Journal of European Public Policy*, *27*(1), 1–19.

Burns, C., & Tobin, P. (2018). The limits of ambitious environmental policy in times of crisis. In C. Adelle, K. Biedenkopf, & D. Torney (Eds.), *European Union external environmental policy: Rules, regulation and governance beyond borders* (pp. 319–336). Palgrave macmillan.

Burns, C., & Tobin, P. (2020). Crisis, climate change and comitology: Policy dismantling via the backdoor? *Journal of Common Market Studies*, *58*(3), 527–544.

Corthaut, T., & Van Eeckhoutte, D. (2012). Legal aspects of EU participation in global environmental governance under the UN umbrella. In J. Wouters, H. Bruyninckx, S. Basu, & S. Schunz (Eds.), *The European union and multilateral governance: Assessing EU participation in United Nations human rights and environmental fora* (pp. 145–170). Palgrave Macmillan.

Delreux, T. (2013). The EU as an actor in global environmental politics. In A. Jordan & C. Adelle (Eds.), *Environmental policy in the EU : Actors, institutions and processes* (pp. 287–305). Routledge.

Delreux, T. (2014). EU actorness, cohesiveness and effectiveness in environmental affairs. *Journal of European Public Policy*, *21*(7), 1017–1032.

Delreux, T., & Happaerts, S. (2016). *Environmental policy and politics in the European Union*. Macmillan International Higher Education.

Delreux, T., & Pipart, F. (2021). Ego vs Alter: Internal and external perceptions of the EU's role in global environmental negotiations. *Journal of Common Market Studies*, *59*(5), 1284–1302.

Downie, D., & Templeton, J. (2016). Persistent organic pollutants. In P. G. Harris (Ed.), *Routledge handbook of global environmental politics* (pp. 411–426). Routledge.

General Secretariat of the Council. (2017). *Thirteenth meeting of the Conference of the Parties to the Basel Convention on the Control of Transboundary Movements of Hazardous Wastes and Their Disposal (Geneva, 24 April–5 May 2017) Eighth meeting of the Conference of the Parties to the Rotterdam Convention on the Prior Informed Consent Procedure for Certain Hazardous Chemicals and Pesticides in International Trade (Geneva, 24 April–5 May 2017) Eighth meeting of the Conference of the Parties to the Stockholm Convention on Persistent Organic Pollutants (Geneva, 24 April–5 May 2017) - Compilation of statements*. Brussels: Council of the European Union.

Gravey, V., & Moore, B. (2018). Full steam ahead or dead in the water? European Union environmental policy after the economic crisis. In C. Burns, P. Tobin, & S. Sewerin (Eds.), *The impact of the economic crisis on European environmental policy* (pp. 19–42). Oxford University Press.

Groen, L. (2018). Explaining European Union effectiveness (goal achievement) in the convention on biological diversity: The importance of diplomatic engagement. *International Environmental Agreements: Politics, Law and Economics, 19*(1), 69–87.

Groen, L., Niemann, A., & Oberthür, S. (2012). The EU as a global leader? The Copenhagen and Cancun UN climate change negotiations. *Journal of Contemporary European Research, 8*(2), 173–191.

Hoerber, T., Kurze, K., & Kuenzer, J. (2022). Towards ego-ecology? How domestic demands challenge the European environmental conscience and EU politics. In T. Hoerber & G. Weber (Eds.), *The European environmental conscience in EU politics. A developing ideology* (pp. 75–94). Routledge.

Hoerber, T., & Weber, G. (2022). Introduction – The European environmental conscience in EU politics. In T. Hoerber & G. Weber (Eds.), *The European environmental conscience in EU politics. A developing ideology* (pp. 1–23). Routledge.

Jansen, K., & Dubois, M. (2014). Global pesticide governance by disclosure: Prior informed consent and the Rotterdam convention. In A. Gupta & M. Mason (Eds.), *Transparency in global environmental governance: Critical perspectives* (pp. 107–131). Massachusetts Institute of Technology.

Kelemen, R. D. (2010). Globalizing European Union environmental policy. *Journal of European Public Policy, 17*(3), 335–349.

Keukeleire, S., & Delreux, T. (2014). *The foreign policy of the European Union* (2nd ed.). Palgrave Macmillan.

Laatikainen, K. (2010). Multilateral leadership at the UN after the Lisbon treaty. *European Foreign Affairs Review, 15*, 475–493.

Liefferink, D., & Wurzel, R. K. (2017). Environmental leaders and pioneers: Agents of change? *Journal of European Public Policy, 24*(7), 951–968.

Oberthür, S., & Dupont, C. (2021). The European Union's international climate leadership: towards a grand climate strategy? *Journal of European Public Policy, 28*(7), 1095–1114.

Oberthür, S., & Groen, L. (2015). The effectiveness dimension of the EU's performance in international institutions: Toward a more comprehensive assessment framework. *Journal of Common Market Studies, 53*(6), 1319–1335.

Oberthür, S., & Groen, L. (2017). The European Union and the Paris agreement: Leader, mediator, or bystander? *Wiley Interdisciplinary Reviews: Climate Change, 8*(1).

Oberthür, S., & Groen, L. (2018). Explaining goal achievement in international negotiations: The EU and the Paris Agreement on climate change. *Journal of European Public Policy, 25*(5), 708–727.

Ohler, F., & Delreux, T. (2021). Role perceptions in global environmental negotiations: From reformist leaders to conservative bystanders. *The International Spectator, 56*(3), 7–23.

Parker, C. F., Karlsson, C., & Hjerpe, M. (2017). Assessing the European Union's global climate change leadership: From Copenhagen to the Paris Agreement. *Journal of European Integration, 39*(2), 239–252.

Pipart, F. (2022). Explaining EU influence in international environmental negotiations. *Journal of European Integration, 44*(8), 1059–1074. https://doi.org/10.1080/07036337. 2022.2052056

Secretariat of the Basel Convention. (2018). *Basel Convention on the control of transboundary movements of hazardous wastes and their disposal.* United Nations Environment Programme.

Vogler, J. (2005). The European contribution to global environmental governance. *International Affairs, 81*(4), 835–850.

Wurzel, R. K. W., Connelly, J., & Liefferink, D. (2017). *The European Union in international climate change politics: Still taking a lead?* Routledge.

2 Climate Change

New Challenges and Unexpected Threats – the Iraqi Case

Paiman Ahmad, Muhammad Usman, and Alhamzah Alnoor

Introduction

In December 2019, the European Commission introduced the European Green Deal (EGD), a policy package that aims to reach climate neutrality by 2050. Climate change and energy transition are aligned with the green economic approach of reducing emissions and meeting the set of targets by 2050. This call for action is aligned with the Intergovernmental Panel on Climate Change and the global concerns regarding global warming beyond 1.5°C, which indicates that artificial action has led to higher temperatures and extreme weather conditions that left adverse effects on biodiversity and ecosystems worldwide (Teske, 2019). Most Middle Eastern fossil fuel producers are exposed to climate change challenges; as a resource-abundant region, the Middle East lacks proper infrastructure, technology, sufficient water resources, strong institutions, and skilled human resources to combat the impacts of climate change, and if necessary, actions are further delayed. Meanwhile, small climate-friendly initiatives have been practiced, such as the case of Egypt's solar panel development, which supports the green transition and shows the government's commitment to renewable energy expansion through huge investments gradually (Farraj, 2024).

Faced with the challenges, climate change will severely affect Iraq's socio-economic well-being; therefore, it is time to consider "the accepted wisdom" while dealing with the government's practical ability to respond to the threats and challenges of climate change (Barnes and Hoerber, 2013). It is in this context that "the European Environmental Conscience" strategically envisions the domestic/local demands and needs driven by the environmental policies of the member states. Meanwhile, this has not prevented EU leadership in global climate initiatives and providing global conscience concerning the current climate challenges. The Green Deal "enables the EU to play a perennial role in becoming a model for sustainable prosperity across the globe" (Dekeyrel et al., 2024). When compared to the European Environmental Conscience consumer behaviour and the country's framework for "transition readiness" with political, economic, and social structures to ensure a secure, reliable, inclusive, and sustainable (World Economic Forum, 2018).

DOI: 10.4324/9781032656359-4

The lack of environmental conscience in Iraq has been the main reason for effective environmental policies contrary to Europe so far. Europe has been the leading player in reforms in climate change policy and adapting it to its foreign policy framework to create globally shared responsibility based on the European environmental vision, which can be reflected in national environmental policies. Since the early 1970s, EU has worked on two main issues concerning the environment and diversifying energy supply, the EU has pursued an international regime on climate change to tackle global climate issues at home and abroad. In this regard, the EGD commits the EU to its role in promoting global climate commitment abroad. However, taking the main responsibility for reducing EGD effects on a regional and international level requires the EU's huge investment in reducing the distributional effects of climate policy, supporting low-income households, and counterbalancing such effects on fossil fuel producers' especially fragile and less diversified economies.

Moreover, in the context of environmental consciousness, the fossil fuel producers have to realise that Europe's contribution to inequalities from climate policies is limited and contributions are restricted to different scales. Positioning the EU for taking global responsibility is yet restricted by regional and domestic issues the region encounters including decision-making, member state interests, and dominant regional thinking in policy-making terms. To achieve EGD goals, the EU policymakers should take action on a global basis, not just a regional one, while joint effort is necessary to reduce the negative effects of EGD and adopt the necessary changes in economic and environmental aspects. Altogether, fossil fuel economies need to adjust their economies to deal with the effects of green transition and adapt to the short-term losses to compensate and respond to crises with the EGD's long-term objectives of climate neutrality and prosperity. It should be noted that building a consensus between the EU and fossil fuel producers based on climate and economic objectives is not readily achievable. The environmental consciousness is overridden by the national and societal economic objectives; moreover, lack of consciousness is the societal understanding of the fact that fossil fuel commodities drive major global industries (Barbieri et al., 2021; Hoeber et al., 2020; Dekeyrel et al., 2024). In addition to serious skepticism, it is important to identify the significance of adaptability and flexibility for economic resilience and indicate the institutions play the role. Taken together, climate resilience and adaptability support fossil fuel economies to adjust to new circumstances created by the climate-constrained world, without severe economic disruptions. It should be noted that energy transition may not provide a just transition for all communities alike thus, the benefits of energy transition in terms of job creation, and access to low-carbon and advanced technologies only go to some countries. To compensate for the burdens of energy transition, and create potential benefits in the long run, for the fossil fuel economies, investing in programmes for training the workforce, economic diversification, and job development is required (Ansari & Holz, 2020; Carley & Konisky, 2020).

As noted, energy transition is sceptical for the fossil fuel-rich countries, as Young (2023), in his article "The Gulf Goes Green", asks if the fossil fuel giants may lead the energy transition, as the Gulf States are concerned about global clean energy transition, highlighting the great transition in terms of power production in the Arabian Peninsula for aiming to lower carbon emission. The COP28 left the Gulf governments with three EU requests: tripling renewables rollout by 2023, better energy efficiency, and accelerating fossil fuel phase-out with a "residual" role for carbon-abating technologies (Civillini, 2023). The COP28 focus on phasing down fossil fuel before 2050 is seen as a region-to-region commitment to the EGD and a net-zero economy as a global responsibility for combating climate change. In all COPs, there has been a unified call for phasing down unabated fossil fuels and related subsidies while envisioning natural gas's vital role during this transitional phase of the energy transition. However, based on Climate Action Tracker's assessment of the big fossil fuel producers, serious actions have yet to be taken to eliminate fossil fuel subsidies, and there is no commitment to ending new investments in oil and gas production.

The green transition has economic, social, and environmental implications for fossil fuel producers. The social impacts are going to be the first affecting certain groups of people who lack necessary social protections. Therefore, in the first phases of the transition governments should enable economic reforms to mitigate the negative impacts of green transition (Gass et al., 2021). Notably, most countries lack access to clean energy, energy poverty is still a problem, and affording and developing renewable energy is a serious concern. On the other side of the story is the Western oil companies who, have paid USD 110bn in dividends and share repurchases. This is more than the global climate finance target of the Paris Agreement, which is estimated to be USD 100bn by 2020 (Climate Action Tracker, 2023).

The story of fossil fuel phase-out is centred on the successful implementation of the EGD, which is supposed to diminish the power of oil and gas in Europe. The success of EGD means the decline of oil prices based on a fall in demand; the EGD leads to the creation of future economies depending on renewable energy resources and work on reducing greenhouse gas emissions to meet the emission targets set in the EGD.

As is expected, the EGD will profoundly influence the fossil fuel-producing economies. Meanwhile, most fossil fuel-producing economies must be equipped and prepared to manage the impact and slow the repercussions. The degree of influence will be contextual and situational based, as each fossil fuel-producing country has a specific condition, which makes the cases different. However, the EGD effects are prevalent and unstoppable as the nature of the transition in Europe will have adverse impacts directly on fossil fuel producers in the Middle East and North Africa.

Concerning the EGD, the implications vary, as it weakens import dependency, tackles global warming at home and abroad, and changes price volatility. Over the years, in fossil fuel economies, environmental issues have gained no importance; that is to say, most MENA producers need to get prepared to reduce the influence of EGD on their economy, people, and community.

Meanwhile, for the EU to meet climate targets and reduce the heavy reliance on Russian fossil fuels, energy diplomacy, and European foreign relations need to be readdressed to minimise Russian influence on the energy supply to Europe until the EU is prepared for a complete energy transition by 2050. However, a one-to-one replacement of former Russian natural gas imports must be fixed for Europe.

Lower energy demand will depress foreign fossil fuel investment, influencing capital shift from advanced economies to the developing World. The EU target is to prepare for a fossil fuel phase-out by 2050, whereas for developing countries, the phase-out means a drastic shift in international financing and investment in oil and gas industries.

Fossil fuel-producing countries are better off if the cost-effect analyses of the EGD are made for each country. Energy transition based on EGD is a state-led commitment to climate neutrality with domestic, regional, and international implications. As noted, the EGD suggests Europe's structural response aligns with a new growth strategy, which firmly sets out the main ambitions for transforming the EU into a resource-efficient and competitive economy (Vela Almeida et al., 2023). Taken from the theoretical debates regarding European Environmental consciences and its effects on other countries, has been reflected in domestic environmental policy-making and government initiatives. Thus, as a case, Iraq has taken the logic of environmental conscience from practical learned lessons of EGD and Agenda 2030, Iraq has started with the National Environmental Strategy and Climate Plan for Iraq (2013–2017); the strategy stresses the consistent challenges Iraq encounters and points out the importance of harmonising and solving problems aligned with national, regional and international environments. Besides, the legal foundation of environmental concern exists in Iraq, including MoE Law No. 37 of 2008, EPI Law No. 27 of 2009, and Article 33 of the 2005 Constitution that stipulates a comprehensive notion that "every individual has the right to live in a safe environment" and the state takes the lead to protect and preserve environment and biological diversity. Alongside the National Environmental Strategy, the National Development Plan (NDP) 2010–2014 focuses on integrating environmental and socioeconomic dimensions as the primary basis of development in Iraq. It should be noted that the needs of the National Environmental Strategy did not meet its goals due to the outbreak of conflict in Iraq against the Islamic State of Iraq and Syria, which has restricted the full implementation of the national strategy. Moreover, the EU and Iraq Partnership and Cooperation Agreement of 2012 emphasised the need to focus on green transition, sustainable development goals, and commitment to global climate change action. To combat the EGD adverse effects, raising awareness regarding European environmental conscience could be aligned with a participatory decision-making process for its applicability to fossil fuel economies (Hoerber, 2013; Fakoussa, 2022).

This chapter deals with the EGD and its implications for fossil fuel economies, taking the case of Iraq as a producing fossil fuel-dependent economy in the Middle East. It explores the importance of EGD and its economic implications for Iraq and analyses the situation of Iraq as a window of hope at the heart of a just transition. The chapter approaches with an exclusive literature review, followed by a discussion of the fossil fuel phase-out and its impact on the Iraqi economy. Moreover, the chapter explores enabling solar energy opportunities in Iraq as a sustainable alternative for incorporating practice into reality. Finally, the chapter concludes with a short discussion of the findings and a policy recommendation concerning EGD and the oil phase-out, as well as its effects on fossil fuel economies such as Iraq.

Literature Review

This chapter addresses Iraq's main challenges as a fossil fuel-producing country because of the implementation of the EGD. Moreover, it highlights the opportunities Iraq can create during and after the fossil fuel phase. While considering the previous studies relevant to this topic, the main perspectives of the EGD aimed at developing a greener economy, which lacks policy and morality for fossil fuel producers. The EGD will change the economic development model for fossil fuel economies and provide opportunities for a gradual transition.

The importance of managing the commons and investing in our future has been addressed in "Our Common Future" in the Brundtland Report 1987; since then, it has become a necessary call for a global agenda for change. Recent global emphasis on climate change, environmental pollution, resource extraction, and climate culture has been addressed by various International Organizations, including UNEP, UNESCO, FAO, IUCN, IMF, and WWF. At the same time, the European Union as a region is a leading actor in coordinating energy transition through different phases and policy packages since the 1950s.

Among the voices concerning energy transition are G7 countries; however, wealthy nations have failed to support poor countries in cutting emissions and coping with climate change (Bocca, 2023). Moving to fossil fuels is a serious action by the supporters of green transition. In contrast, fossil fuel economies still need to prepare and step up the ladder to cope with the energy transition in the near future. To this end, there is no "one-size-fits-all" policy; thus, fossil fuel phase-out has various implications for the producing countries and International Oil Companies. To meet the main goals of the energy transition, meeting the 2015 Paris Agreement on climate change, similar to the Green Deal, needs a drastic reduction in fossil fuel use, which aims to cut emissions and approach a green economy (Muttitt et al., 2016). In considering the EGD, the lack of a just transition is at the core of the concern for the fossil fuel producers because there is no guarantee of a "just transition" during the phase-out era. Therefore, fossil fuel producers will pay the heavy price of the green transition (Rosemberg, 2017). As expected, EGD results in the fossil fuel phase-out, which affects economic opportunities for companies and citizens; thus, energy transition creates disparities in the labour market; if "a just transition" is not secured, hence governments need to manage well and swiftly, the transition to contribute to making decent work for all, ensuring social inclusion and keeping an eye on eradication of poverty, not the opposite (ILO, 2015). During the different phases of the green transition, the vital role of government is to be noted concerning reducing risks, building resilience to the efforts of climate change, and necessary adoption to economic, social, and environmental costs. Meanwhile, as an example in Iraq, Basrah governorate, for combating climate change, it is recommended to educate people to learn and contribute to reducing climate change effects (Salman, 2022).

Based on the IEA, Iraq is the second-largest crude oil producer in OPEC after Saudi Arabia. It holds the World's fifth-largest proven natural oil reserves, at 145 billion barrels, representing 17% of the Middle East and 8% of the global reserves. Additionally, crude oil production grew by 1.7 million barrels per day from 2013 through 2019, and Iraq's production averaged 4.7 million barrels per day in 2019. The effects of the green transition on the Iraqi economy will be massive, as Iraq depends heavily on oil revenues. In 2022, crude oil export revenue accounted for an estimated 92% of Iraq's total government revenues. For a case like Iraq, the energy transition needs a long-term plan in which the employment dimension of climate policy and social dimension for handling the transition from fossil fuel dependence is required to reduce the severe impacts of the green transition (Gass and Echeverria, 2017; World Energy Review, 2023; Rosemberg, 2017). In the context of transition, similar to transition and reduction in coal use, governments must address climate change as a challenge. Thus, governments are required to ensure that energy transitions are people-centred and inclusive. Effects could be opposite to European Economies with advanced technologies and developed economies, as it is estimated that the transition to net-zero can bring new employment opportunities, with 14 million jobs created by 2030 by investing in clean Energy (International Energy Agency, 2021).

Meanwhile, governments with massive fossil fuel resources have not joined the Fossil Fuel Non-Proliferation Treaty initiative, which calls for ending new exploration and production, also focusing on a fair phase-out of existing production and securing a just transition for the fossil fuel workers, communities, and producing countries. The treaty calls for international cooperation to prevent the proliferation of fossil fuels, manage the decline in production, and induce a just and equitable transition. Meanwhile, reducing the transition costs for workers and communities while transitioning to renewable Energy and securing economic diversification. While the EU is at the forefront of the energy transition race, to afford complacency, Europe needs to support fossil fuel economies to catch up with the same journey at different speeds. Otherwise, the transition to renewable Energy will not secure the cutting of required emissions in fossil fuel-dependent economies (Romano, 2023).

Fossil Fuel Phase Out and its Effects on the Iraqi Economy

For the last four decades, Iraq has relied on the fossil fuel sector as the main driver of economic development and growth. Iraq is the third largest exporter of oil and gas after Russia and Saudi Arabia. Meanwhile, OPEC's second-largest crude oil producer, Iraq, is the World's fifth-largest proved natural oil reserves at 145 billion barrels, accounting for 17% of proved reserves in the Middle East and 8% of global reserves.[1] Notably, Iraq's fossil fuel dependence is amplified by the share of fossil fuel exports in Iraq's total exports as fossil fuels dominate export share.

According to World Energy Review, 2023, "In 2019, crude oil export revenue accounted for an estimated 92% of Iraq's total government revenues." This record is significant evidence of the heavy dependence on fossil fuels, which brings various challenges to Iraq. Regardless of the substantial role fossil fuel products play in the Iraqi economy, there is a global and European shift toward a low-carbon economy, which brings severe challenges to the Iraqi economy and community alike.

Fossil fuel phase-out has become a concern for Europe and global supporters of climate action. Meanwhile, large oil and gas producers must be more conscious of the effects of the EGD's success and net-zero economy. In parallel, the oil and gas-producing countries still need to commit to the Paris Agreement for oil and gas production. Meanwhile, the current condition of many oil and gas producers is not convenient for supporting the fossil fuel phase-out, as developed countries also did not provide financial and technical support to develop renewable energy resources to tackle global emissions (Climate Action Tracker, 2023). Align with this; the ethical leadership of the EU is necessary for considering the other countries' policy deficiencies and the need for capacity building for sustainability and long-lasting development (Rutazibwa, 2010). Furthermore, the European environmental policies and strategies may not fit the Middle Eastern fossil fuel economies, but governments can focus on designing green incentive regimes while considering the country's circumstances and vision of future development (Larhlid, 2023). Indeed, country conditions vary, and public opinion about environmental policies is country-based, considering European public perception about the environment is concentrated on a few issues, including decision-making where two-thirds of European citizens believe decisions to be made jointed within the EU rather than individual national governments. There are clear-cut differences in considering how to protect the environment for the EU, in these factors can be important for other countries such as Iraq. Legislative aspects are vital for environmental protection and the awareness of the public and other stakeholders. As suggested the EU should assist the non-EU countries to improve their environmental standards and more investments to be made for environmental protection. Furthermore, Europe needs to push to boost global environmental conscience beyond its borders. Adding to this, Europe needs to collaborate with other countries to assert its global leadership in combating climate change and achieving EGD objectives collectively, particularly by assuring the EU Global Strategy and the European Neighborhood Policy aligned with EGD (European Commission, 2014; Leonard et al., 2021; Bennis, 2021). Concerning green transition, action on climate change is crucial in reducing climate change, especially by decreasing the use of fossil fuels. Public knowledge regarding climate change affects policy implication, adaptation, and public contribution to climate change action. While the data indicates that Iraqis compared to the rest of the European citizens know a little about climate change (International Public Opinion on Climate Change, 2022).

Less progress is noted in developing countries, particularly in fossil fuel-abundant countries; according to Global Electricity Review 2023, the World has abundant supplies of wind and solar, which can be converted to energy resources coefficient and often at a lower cost than fossil fuels. As country conditions are peculiar, energy resources for electricity, heat, and transport, besides industries, need to step into a clean transition and start investing in enabling renewable energy for the community's needs (Motyka, 2023). In light of the EGD, for the MENA heavy fossil fuel producers such as Iraq, the Nationally Determined Contributions (NDCs) country commitments aligned with the Paris Agreement have yet to progress or start. While defining climate risks and impacts is not well decided, the required actions considering the priority sectors that produce heavy emissions must be addressed under the Paris Agreement's terms (UNDP, 2023). Evidence shows that growth in fossil fuel-dependent economies is heavily foreign rent-dependent, which has been at the expense of climate. For years, fossil fuel-based economies have grown at the expense of climate and the environment; the quantity has been the concern rather than the quality of the climate and our shared future. For most oil and gas-producing countries, a setting for combining action on climate and growth needs to be prepared. Therefore, a "decisive transition" to a low-emission, high-growth, and resilient future has yet to be created (OECD, 2017).

In line with this, the EGD threatens the economic foundation of the fossil fuel economies. However, the local pollution due to exploration and extraction has driven momentum for necessary reforms, and environmental hazards pose severe challenges for communities locally and globally. Dealing with climate damages compares it to economic gains. It is estimated that a decisive transition occurs based on a country's economic structure, which, even for fossil fuel exporters, can offset losses and generate economic growth if only policies are appropriate (OECD, 2017). For fossil fuel producers, climate change and adaptation to the changes are urgent, as mitigation and adaptation are outside the policy agenda in the producing countries. To a large extent, uncertainties are associated with adaptation and mitigation strategies of the government to reduce the effects of climate change, especially on small and vulnerable developing economies. Unfortunately, the economic uncertainties and consequences of the energy transition based on climate change initiatives are not well studied and defined by the fossil fuel economies (Bellon, and Massetti, 2022).

The State of Energy Transition and Enabling Solar Energy in Iraq

The green transition and shifting towards a low-carbon economy are not so far integrated into the context of the Iraqi economy. Energy transition is firmly associated with environmental concerns and global warming. Energy transition challenges should be studied more for fossil fuel-dependent econo-mies, especially in the developing World. It is a matter of fact that energy transition positions oil and gas producers at risk, making them vulnerable to a decline in energy demand. The effects and implications are different based on the country's conditions and the nature of the energy transition. However, if EGD fails to engage fossil fuel producers, it will have severe regional and global implications.

It should be noted that EGD orients energy consumption towards climate change to reduce the negative impacts and keep energy transition ambition balanced. At the heart of the energy transition debate is the logic of collective action, which is required to match the public and private interests of the stake-holders involved in general. Considering the repetitive price shocks since the 1970s, large oil producers like Iraq and Saudi Arabia have yet to do anything regarding necessary policy changes, even after the frequent shocks in oil prices. For years, there has been a need for fundamental policy change (Allawi & Birol, 2021; Grundinger, 2017). It can be realised that EGD is not just dealing with Europe alone, but climate diplomacy, readdressing international trade, carbon tariffs, and transnational investment relations accompany it (Teevan et al., 2021). If undertaken wisely, energy transition based on EGD has to be considered a win-win situation for Europe and fossil fuel economies. European Union's leadership in green transition is shown as a geopolitical steward who needs to buy into another region with conflicting interests to create a shared responsibility for safeguarding a better and safer future for all (Rutazibwa, 2010).

For fossil fuel-dependent economies, dealing with energy transition is mul-tifaceted, as the transition to renewables conveys more than it can be esti-mated. For fossil fuel producers, the international oil and gas companies play a vital role in the transition to align with EGD. However, recent literature and data show that there needs to be more consensus among international oil and gas companies concerning their interest in and actual response to renewables. At the heart of this debate is that successful participation in energy transition depends on policy change and political will. In many developing countries, heavy investments in oil and gas exploration and concessions still need to be overcome. Foreign firms with high interest mainly dominate the energy sector. The political difficulty of a systematic change makes it undesirable for the energy producers to step into the energy transition boat. Fossil fuel-dependent economies risk economic development and growth based on finite and nonre-newable commodities, and governments are less concerned about the signifi-cant challenges that energy transition can bring. Another vital aspect of conventional energy resources is the externalities such as price shocks; even though countries such as Iraq, Saudi Arabia, Venezuela, and Russia, are accustomed to external shocks, remarkably prices, the weaknesses of the poli-cies and economic costs of the shocks are not estimated appropriately to over-come the consistent challenges.

Another severe challenge of the energy transition phase is the government's political support function, which can be reflected in the consumers' prices and the profits of energy producers. This is the equation where environmental concerns do not play a serious role, while economic returns do.

The new economies will focus on climate change demands, and geopolitical relations and international cooperation will influence the energy transition era. In the past decades, the World has seen wars to control oil resources, examples of the 1980–1988 Iran–Iraq war, the 1990s Gulf War, the 2003 Iraqi War, the ongoing conflict in Delta–Nigeria, and political tensions in Venezuela. Moreover, there is competition between oil-exporting countries and oil-importing countries, where the power of production has influenced the price, and few historical price shocks in the past decades have been recorded (Alkin and Urpelainen, 2018, Hafner & Tagliapietra, 2020).

Economic transformation based on EGD leads to a fossil fuel phase and a decline in fossil fuel consumption. In this scenario, the European Union must practice Green Deal diplomacy, which can infuse and influence foreign governments to integrate soft regulatory power for climate response. Therefore, the external aspect of the Green Deal and the outcomes to be studied well regarding fossil fuel economies are to enhance renewable energy capacity building and adapt new investments in fossil fuel fields (Leonard et al., 2021). More important is dealing with the investments and assets of the International Oil Companies, which operate in the fossil fuel fields in the Middle East, especially in Iraq after 2003, since cutting emissions seems only possible if they are closed (Hoffman and Ely, 2022). However, the fossil fuel price fluctuation often complicates the energy transition, as cheaper oil and gas demotivate the transition to renewable energy, especially in the producing countries, including Iraq, where price and access are more affordable than in Europe.

Concluding Remarks and Policy Recommendations

In the last 30 years, fossil fuel-dependent economies have faced uncertain oil prices and experienced few price shocks. The three recent price shocks of 2008, 2014, and 2020 have been different as they were recorded due to oil demand, opposite to the 1970s oil crisis due to the Middle Eastern political turmoil. The future of fossil fuel economies is associated with the EGD and the demand scale, as the energy transition in Europe will influence fossil fuel economies directly, and a decline in demand for fossil fuel commodities is highly expected. Align with this, the process of complete energy transition takes time, according to the projections made in the EGD, to replace oil with natural gas as a more climate-friendly commodity. Thus, the potential for change can speed up the economic recovery in the Middle East and North Africa (MENA) region if the political will exists. Based on facts, green transition takes time; thus, fossil fuel economies can plan to adopt EGD effects as serious change for oil and gas takes place between 2030 and 2050 (Leonard et al., 2021).

The launch of the EGD influences the EU's external relations with fossil fuel producers, especially the countries exporting to Europe. In comparison,

EGD redefines external diplomacy based on the low-carbon transition to reduce the Green Deal's economic impacts on fossil fuel-dependent economies such as Iraq.

At the glance of EGD, Europe needs to take the lead in settling the conflicting priorities for fossil fuel producers in the Middle East and other parts of the World if the main target of EGD is cutting emissions and achieving a sustainable green transition.

To avoid severe economic costs, fossil fuel governments must deal with EGD implications based on a just transition and sustainable investment in diversifying their economies. Radical action is required from fossil fuel producers to avoid regional and global economic damage caused by climate change, which is estimated to be $ 22.5 trillion by 2100. As fossil fuel national securities are associated with their socio-economic conditions, the government's ability to deal with and respond to climate threats is vital. The national pressures can influence international and regional security if climate change leads to mass population displacement, migration, and political unrest. The current climate conditions in southern Iraq have been seen as recent pressures due to the decrease and shortage in water resources. From this perspective, the government must embed the "environmental conscience" in energy policy as a moral obligation and shared responsibility. Besides, the fiscal policy should fuel sustainability and reduce government dependence on fossil fuels by investing in economic diversification and adjusting economic recovery based on sustainability wisdom (Barnes and Hoerber, 2013; IMF, 2021).

Regardless of the challenges and risks the EGD brings to the Iraqi economy, the low-carbon transition provides Iraq with opportunities such as diversifying its economy, developing new sectors such as religious tourism, gradually reducing fossil fuel dependence, and investing in renewable energy resources. Examining Iraq's economic conditions shows that Iraq's dependence on fossil fuels currently cannot support sustainable growth and economic development. Besides, this study suggests an economic reform aligned with the EGD policy package, which can tackle the potential challenges and reduce the adverse effects of energy transition.

As a fossil fuel-dependent country, Iraq needs to prepare for policy reforms and prepare the Iraqi economy to become more resilient to climate change threats. Meanwhile, strategically investing in economic diversification is essential for reducing fossil fuel dependence, shifting to renewable energy resources, and developing other sectors of the economy. To this end, Iraq has potent solar power, similar to the rest of Middle Eastern and African countries. Therefore, investment in massive solar power plants in Iraq and MENA can position the region as the World's largest solar power producer.

To this end, the energy transition in the EGD picture is still early, which indicates different implications and influences across the region and extends to fossil fuel producers. Fossil fuel producers must address the risks, threats, and challenges in their economic policies. Overall, energy transition "greening" must be central in Iraqi policy-making to stimulate the practical application of required changes in the country's economic culture (Solorio et al., 2013). Moreover, for European climate initiatives to fit the interest of EU partners, one-size-fits-all is not applicable, the EU needs to adopt specific approaches that fit the context of each partner while considering the domestic competitive advantage as a priority (Leonard et al., 2021).

Note

1 Iraq | HKTDC Belt and Road Portal.

Bibliography

Alkin, M., & Urpelainen, J. (2018). *Renewables, the politics of a global energy transition, Cambridge, the United States of America*. MIT Press.

Allawi, A., & Birol, F. (2021). Without help from oil-producing countri/es, net zero by 2050 is a distant dream, according to the OECD. Without help from oil-producing countries, net zero by 2050 is a distant dream – Development Matters (OECD-development-matters.org).

Ansari, D., & Holz, F. (2020). Between stranded assets and green transformation: fossil-fuel producing countries towards 2055, *World Development*, *130*, 104947.

Barbieri et al. (2021, December). Sustainability transition and the European green deal: A macro-dynamic perspective, Eionet Report-ETC/WMGE, European Topic Center on Waste and Materials in a Green Economy.

Barnes, P. M., & Hoerber, T. C. (2013). *Introduction: Establishing the research questions and methodological framework*. Routledge.

Bellon, M., & Massetti, E. (2022). *Economic principles for integrating adaption to climate change into fiscal policy*. Staff Climate Notes, IMF.

Bennis, A. (2021, January). *Power surge: How the European green deal can succeed in Morocco and Tunisia*, European Council on Foreign Relations, Policy Brief.

Bocca, R. (2023). Energy transition, SDG 07: Affordable and clean energy. *World Economic Forum*.

Carley, D., & Konisky, D. M. (2020). The justice and equity implications of the clean energy transition, *Review Article, Nature Energy*. https://doi.org/10.1038/s41560-020-0641-6

Civillini, M. (2023, July). EU to push for fossil fuel phase-out 'well ahead of 20250' at COP28. *Climate Home News*. https://climatechangenews.com/2023/07/12/eu-fossil-fuel-phaseout-2050-cop28/?fbclid=IwAR1iHx5d8buvIfqIUJIlLeL42GFLeNbRA1v XwIUc0QTyFdC_Mkg_bOoB2JM

Climate Action Tracker. (2023, June). Countdown to COP28: Time for the World to focus on oil and gas phase-out; renewables target distractions like CCS. *New Climate Institute*.

Dekeyrel et al., (2024). The Green Deal in times of poly crisis: Aligning short-term responses with long-term commitments, Sustainable Prosperity for Europe Programme, Discussion paper, European Policy Center.

EIA. (2017). Today in Energy, September 14, available at: https://www.eia.gov/todayinenergy/detail.php?id=32912

EIA. (2022). Iraq-Analysis-energy sector highlights, February 14, available at: https://www.eia.gov/international/overview/country/irq

European Commission. (2014). *Attitudes of European citizens towards the environment – Report*, Publications Office. https://data.europa.eu/doi/10.2779/25662

Fakoussa, D. (2022). The EU green deal, a new momentum for democratic governance in the MENA region? EUROMESCO Paper, No. 47.

Farraj, Y. A. (2024, April). Egypt to build 2 solar power stations with $20.6 million in EU funding, Middle East Economy. https://economymiddleeast.com/news/egypt-to-build-2-solar-power-stations-with-20-6-million-in-eu-funding

Fossil Fuel Non-Proliferation Treaty. (2020). Briefing Note: Aligning fossil fuel production with 1.5°C and the Paris agreement. https://fossilfueltreaty.org

Gass, P., & Echeverria, D. (2017). Fossil fuel subsidy reform and the just transition, GSI report, International Institute for Sustainable Development.

Gass, P., Gerasimchuk, I., Kuehl, J., Roth, J., & Wooders, P. (2021). *Just transition to a green economy, Sector Protect 'employment promotion in development cooperation* (p.15). International Institute for Sustainable Development (IISD/).

Grundinger, W. (2017). *Drivers of energy transition*. Springer.

Hadji-Lazaro et al. (2022). Taking the temperature of the European green deal. FEPS Policy Study, European Political Foundation/Karl Renner Institute.

Hafner, M., & Tagliapietra, S. (2020). *The geopolitics of the global energy transition, Lecture Notes in Energy* (Vol. 73). Springer.

Hoerber, T. (2013). *The origins of energy and environmental policy in Europe the beginnings of a European environmental conscience* (1st ed.), Routledge.

Hoerber, T., Weber, G., & Cabras, I. (2020). The role of sustainability as ideology in shaping collaborative governance within the European Union, *International Public Management Review*, *20*(2), 25–39.

Hoffman, A. J. & Ely, M. D. (2022). Time to put the fossil-fuel industry into hospice, *Stanford Social Innovation Review*, Fall, Vol. 29 pp.29–37.

International Energy Agency. (2021). Net zero by 2050: A roadmap for the global energy sector. International Energy Agency, Special Report.

International Labour Organization. (2015). Guidelines for a just transition towards environmentally sustainable economies and societies for all. ILO. http://www.ilo.org/global/topics/green-jobs/publications/WCMS_432859/lang--en/index.htm

International Monetary Fund, (2021). Iraq, 2020 Article IV constitution, IMF Country Report No. 21/38, Washington. D.C. IMF.

International Public Opinion on Climate Change, 2022). Yale program on climate change communication, Yale School of the Environment, Meta.

Larhlid, A. (2023, September). *Is a green deal the Middle East's path to sustainability?* PWC. https://www.pwc.com/m1/en/publications/an-incentive-to-action-european-green-deal-and-the-middle-east.html

Leonard, M., Pisani, J.F., Shapiro, J., Tagliapietra, S., & Wolff, G. (2021). *The geopolitics of the European green deal*, Policy, Contribution 04, Bruegel.

Motyka, M. W. (2023, April). *Global electricity review*. Ember. https://ember-climate.org/insights/research/global-electricity-review-2023

Muttitt, G., McKinnon, H., Stockman, L., Kretzmann, S., Scott, A. and Turnbull, D. (2016). *The sky's limit: Why the Paris climate goals require a managed decline of fossil fuel production*. Oil Change International. http://priceofoil.org/2016/09/22/the-skys-limit-report

OECD, (2017). *Investing in climate, investing in growth*. OECD Publishing.

Roggema, R. (2009). *Adaptation to climate change: A spatial challenge*. Springer.

Romano, V. (2023, April). *The end of the Fossil Age has begun*. ERUACTIV. https://www.euractiv.com/section/energy-environment/news/end-of-fossil-age-has-begun-analysts-say

Rosemberg, A. (2017). *Strengthening just transition policies in international climate governance*. The Stanley Foundation. https://www.stanleyfoundation.org/resources.cfm?id=1629

Rutazibwa, O. U. (2010). The problematics of the EU's ethical (self)image in Africa: The EU as an 'ethical intervener' and the 2007 joint Africa–EU strategy. *Journal of Contemporary European Studies*, 18(2), 209–228.

Salman, N. A. (2022). *Climated-induced displacement efforts in Basrah and justice solutions for community resilience*. Cordaid.

Solorio, I., Bechberger, M., & Popartan, L. (2013). *The European energy policy and its 'green dimension'*. Routledge.

Teske, S. (2019). *Achieving the Paris climate agreement goals*. Springer Nature.

Teevan, C., Medinilla, A., & Sergejeff, K. (2021). The green deal in EU foreign and development policy, Making Policies Work, BRIEFING NOTE NO. 131, the European Centre for Development Policy Management (ECDPM), May.

UNDP. (2023, May). What are the NDCs, and how do they drive climate action? https:// climatepromise. undp. org/ news- and- stories/ NDCs- nationally- determined- contributions- climate- change- what- you- need- to- know#:~: text=NDCs% 20represent%20short-%20to%20medium-term%20targets%20and%20typically% 20include,closer%20to%201.5°C

Vela Almeida, D., Kolinjivadi, V., Ferrando, T., Roy, B., Herrera, H., Goncalves, M. V., Van Hecken, G. (2023). The greening of empire: the European green deal as the EU first agenda, *Political Geography*, 105, 102925.

World Energy Review. (2023). 22nd Edition of World Energy Review, Roma, Italy, ENI, avaibale online at: https://www.eni.com/en-IT/media/press-release/2023/10/eni-publishes-22nd-edition-world-energy-review.html

World Economic Forum. (2018). Fostering effective energy transition a fact-based framework to support decision-making. Insight Report.

Young, K. E. (2023). The Gulf goes green; can the fossil fuel giants lead the energy transition? *Foreign Policy*. https://www.foreignaffairs.com/persian-gulf/gulf-goes-green-fossil-fuel-energy-transition

The Growing Role of Green Finance and Sustainability-Oriented Financial Regulation in the European Union

3 Financing Europe's Transition to a Low-Carbon Economy

The Role of Sustainable Finance

Chiara Pappalardo

Introduction

Sustainable finance is an investment strategy and a new area of policy concern for the European Union (EU). As an investment strategy, sustainable finance can be traced back to the socially responsible investment movement of the 1980s–1990s initiated by churches, unit trusts and mutual funds (Sparkes, 2006; Sparkes & Cowton, 2004). Initially of minor economic importance, by the early 2000s setting ethical parameters to investment portfolios matured into a philosophy adopted by a growing proportion of large investment institutions, corporate executives could no longer ignore (Sparkes, 2006). This type of investment approach has grown strongly in Europe, the United States and China. Today, sustainability-oriented funds represent a sizeable portion of the global financial market.

As a new area of EU policy concern, sustainable finance reflects discrete political and regulatory choices. The European Commission (Commission) defines sustainable finance as "the process of taking environmental, social and governance (ESG) considerations into account when making investment decisions in the financial sector, leading to more long-term investments in sustainable economic activities and projects". Sustainable finance plays a critical role in advancing the economic goals set forth in the European Green Deal (Commission, 2019). Within the overarching objective of reaching carbon neutrality by 2050, sustainable finance is conceived by EU institutions as a tool to bridge the climate finance gap. A related objective of the EU sustainable finance policy is to address short-termism in finance to improve market stability during the transition to a low carbon, more resource efficient and sustainable economy.

One stream of scholarship investigates whether a common environmental conscience exists in the EU, and whether it serves as a source for further European integration in the form of spillovers to other policy sectors (Hoerber, 2012; Hoerber & Weber, 2022). Sustainable finance is a field in evolution and, therefore, subject to changes and policy adjustments in the coming years. With this in mind, this chapter asks whether sustainable finance policy fits the

DOI: 10.4324/9781032656359-6

environmental conscience paradigm. In the chapter, I examine how initiatives align with and support both energy and environmental policies. I highlight the innovative character of some of these initiatives together with its main drivers. I also explore how the expertise of the Commission has contributed to shape the sustainable finance landscape. I conclude that the European environmental conscience spill over to the field of financial regulation is expected to further the integration of EU capital markets towards sustainability. However, challenges remain within the current regulatory framework to support a truly sustainable economy.

Sustainable Finance as a New Area of EU Policy Concern

Until recently, sustainability was not part of the debate on financial regulation (Ahlström & Monciardini, 2022). Traditionally, the field of finance has been concerned with profit maximisation. The prevalent view was that integrating social and environmental considerations in investment decisions could hurt financial returns. Several global financial crises in the early 2000s laid bare the limits of a *laissez-faire* approach to financial regulation (Weber, 2015). Progressive disillusionment with the financial industry called for stronger government intervention. At the same time, climate change was coming to the fore of the policy debate globally. Citizens' discontent in this space began to be directed towards governments and businesses both seen as not doing enough to address the growing impacts of global warming (Tendero, 2021). Sustainable finance emerged as a means to address these twin crises: the financial industry's legitimacy and the need to invest in innovation and technology to reduce greenhouse gas (GHG) emissions. As an investment strategy through which people "vote with their wallet" and as a means to raise the resources necessary to foster public and private innovation in alternative energy sources, sustainable finance offers decision makers a new political economy narrative, one where the finance industry can be put to work to serve the common good and play a positive role in society.

Beginning in the 2000s, we observe an increased focus and activity towards sustainability in the field of finance. As discussed in chapter 4, there is a vast universe of tools used by EU institutions to promote sustainable development objectives. Prior engagement with financial institutions, businesses and civil society is the best path for lawmakers and regulators to ensure that regulations are fit for purpose and to increase the probability of their compliance. However, when a multiplicity of actors gain a seat at the decision-making table tensions also emerge. Specifically with regards to reporting obligations the Commission launched its first multistakeholder consultation process on corporate social responsibility in 2001 to identify with industry and civil society approaches, tools and practices that were credible and coherent enough and could be independently verified (De Schutter, 2008, pp. 207–208). The goal was to create a set of commonly agreed standardised and certifiable benchmarks and reporting requirements to help financial markets reward corporate sustainability best practices and sanction the worse forms of behaviour (De Schutter, 2008, p. 216). However, business groups successfully persuaded the Commission that such an approach would hinder innovative, bottom up, market-driven solutions and ultimately damage EU competitiveness (De Schutter, 2008, p. 213). In the final Communiqué, the Commission declined to take any action for fear that imposing new obligations and administrative requirements on European businesses would be counterproductive. The position was in direct contradiction with the Commission's initial idea that ESG-enhancing policies would not only support the advancement of public goals but also be good for business (De Schutter, 2008, pp. 215–217).

A shift in policy emerges in the aftermath of the 2008 Global Financial Crisis. In 2011, the Commission acknowledged for the first time that certain regulatory measures create an environment more conducive to businesses voluntarily meeting their social responsibility (European Commission, 2011, p. 3). One objective was to align the European and international visions on corporate social responsibility (European Parliament, 2020, pp. 21–22). Internationally, pressure was mounting to hold businesses accountable for their adverse impacts on people and the planet, culminating in the endorsement by the United Nations Human Rights Council of a set of guiding principles on business and human rights (Ruggie, 2011). The principles, while not binding, laid down the essential requirements for corporate respect of human rights, and inspired several member states' legal frameworks on corporate due diligence (European Parliament, 2020, p. 17). Between 2006 and 2013, the United Kingdom, Sweden, Spain, Denmark and France had introduced disclosure requirements that went beyond the narrow EU framework, creating the possibility of additional compliance costs for companies operating in more than one member state (European Commission, 2014, p. 1; Nadakavukaren-Schefer, 2020, pp. 86–90). A related objective was to ensure that horizontal (i.e., across businesses regardless of the type of products or services they provide) and sector-specific approaches to implementation in the member states would be sufficiently harmonised (European Parliament, 2020, p. 21).

Yet, at its core, corporate social responsibility and its methods of reporting remained conceptually understood, and legislatively framed, as a series of voluntary actions that companies undertake to address social and environmental concerns in their interaction with stakeholders beyond their legal obligations. With Directive 2013/34/EU, large companies and public interest entities were encouraged to make publicly available information regarding their performance on non-financial matters including environmental and employee matters. This information had to be provided "where appropriate" and "to the extent necessary for an understanding of the undertaking's development, performance or position" (article 19, paragraph 1). It was to be included in the management report of the company to give "a fair view of the development and performance of the undertaking's business and position together with a description of the principal risks and uncertainties that it faces". Simply put, the Directive did not require companies to implement ESG-promoting policies. The only requirement was to note the reasons for the lack thereof (article 20, paragraph 1(b)). Companies also enjoyed wide discretion on how to report on those matters. As a result, fewer than 10 percent of the largest companies (less than 2,500) were disclosing non-financial information regularly (European Commission, 2014, p. 1).

Recognising the limits of voluntary disclosures and business self-regulation, the Commission decided to propose revisions to the non-financial reporting provisions (European Commission, 2014, p. 3). Under Directive 2014/95/EU, Article 1, "large undertakings which are public interest entities exceeding on average 500 employees during the financial year *shall* include a non-financial statement in their management report" (emphasis added). The Directive raised the threshold from 250 to 500 employees, therefore, reducing the scope of application compared to Directive 2013/34/EU. Nevertheless, it covered approximately 6,000 large companies and groups across the EU (European Commission, 2014, p. 2), a number three times higher than the number of companies already reporting on ESG factors. "Public interest entities" included listed companies as well as some unlisted companies such as banks, insurance companies and other companies that are so designated by member states because of their activities, size or number of employees (European Commission, 2014, p. 2). These companies had to publish information on environmental, social and employee matters, respect for human rights, anti-corruption and bribery matters (article 1). In disclosing such matters, they could pick their preferred method of reporting (e.g. the United Nations Global Compact, ISO 26000, the German Sustainability Code) whichever meshed best with their own characteristics or business environment (Directive 2014/95/EU preamble, paragraph 9; European Commission, 2014, p. 1). Companies could, therefore, continue to use any international, European or national guideline, including their own methodology and criteria, further jeopardising comparability rather than improving it. They had to meet their reporting requirements for the first time in 2018, for information relating to the fiscal year 2017.

Furthermore, between 2017 and 2019, the Commission issued two sets of guidelines on methodology for reporting non-financial information, comprising a list of non-financial key performance indicators, both general and sectoral, to help companies best communicate their policies and initiatives on environmental and social matters. The Commission chose, however, not to make these guidelines binding (Directive 2014/95/EU, article 2). One concern was to continue to allow for maximum reporting flexibility to avoid stifling innovation in reporting practices and keeping the administrative burden for companies at a minimum. Use of multiple reporting standards and methodologies made non-financial information difficult (if not impossible) for investors and other stakeholders to compare, and for member states to enforce meaningfully and monitor trends and patterns. On the one hand, Directive 2014/95/EU required certain large companies to report on non-financial matters listed in the Directive itself; on the other hand, these companies could avoid reporting altogether by choosing to explain instead why they would not pursue policies in relation to one or more of those matters (article 1, "comply or explain"). Moreover, while the Commission undertook the task to formulate and provide reporting guidelines in line with EU international obligations and international bodies recommendations, it left it to the discretion of each individual company whether to use them or not. Only a small set (20 percent) of large companies covered by Directive 2014/95/EU were ultimately applying reporting standards (European Commission, 2021b).

The European Green Deal marks a second policy shift involving sustainability in finance. Huge investments are needed to transform the economy towards sustainability and reach carbon neutrality by 2050 (Tendero & Weber, 2024, p. 3). Since the scale of the investment need is beyond the capacity of public budgets alone, most investments and R&D expenditures will have to come from the private sector (European Commission, 2021a). Corporate investment is already the main driver of economic growth in most European countries (Tendero & Weber, 2024, p. 4). Under the EU's new green growth agenda, reporting requirements play a critical role in directing capital investments towards sustainable economic activities and projects. As Weber and Tendero note in their introductory chapter, financial regulation has been part of the process of EU integration from the beginning. The underlying need to integrate member states, financial markets is foundational to the EU's ability to build a common market and guarantee the free movement of capital in the Union. More developed and integrated capital markets are crucial to unlock investments for companies, especially small and medium size enterprises (SMEs), and to provide funding for innovative projects in the context of transitioning to a sustainable economy (European Council, 2019, p. 3). Moreover, the mispricing of climate risks poses a growing threat to the stability of the global financial system with consequences potentially similar to that of another major financial crisis (European Central Bank, 2022). The ability of banks and insurance companies to identify and manage sustainability risks and absorb financial losses arising from them requires amendments to the European prudential framework to ensure ESG factors are consistently included in risk management systems and by supervisors of credit institutions and investment firms (European Commission, 2021a, pp. 12–13; European Commission, 2024). Finally, greater transparency and credibility of both ESG funds and ESG rating agencies also requires a regulatory response at the EU level (European Commission, 2014; European Commission, 2023b) to harmonise member states' approaches to sustainable investing and combat greenwashing effectively.

Addressing the EU's Climate Finance Gap

With the adoption of the European Green Deal, strengthening the sustainable finance framework becomes instrumental to achieve the EU economic goals. In 2018, the Commission first published a "Sustainable Finance Action Plan" recognising that environmental and social factors are often not sufficiently taken into account in investment decisions since such risks tend to materialise over a longer term horizon (European Commission, 2018, pp. 2–3). The 2018 Plan sets forth three overarching objectives: (1) Reorient capital flows towards sustainable investment to achieve greener and more inclusive growth; (2) Manage financial risks stemming from climate change, resource depletion, environmental degradation and social issues; (3) Foster transparency and long termism in financial and economic activity. The plan aims at better connecting the financial sector with the specific needs of the European economy. The economic needs outlined in the European Green Deal include mobilising private capital for energy efficiency, renewable energy and green infrastructure projects as well as better manage and integrate environmental and climate risks in the financial system to avoid stranded assets (European Commission, 2019, pp. 16–17). In the plan, the Commission directly acknowledges that taking longer-term sustainability interests into consideration makes economic sense and does not necessarily lead to lower returns for investors. The Commission also notes that, going forward, it will be necessary to strike an appropriate balance between flexibility and the standardisation of disclosures to generate high quality and sufficient data for investment decisions.

However, with the subsequent pandemic and economic downturn, the global landscape had changed and a new look at the role of sustainable finance within the EU economic agenda appeared necessary. As part of a sustainable recovery from the Covid-19 crisis, in 2021 the Commission published a "Strategy for Financing the Transition to a Sustainable Economy" further emphasising the key role that the private sector plays in financing the transition by reorienting capital flows towards sustainable funds. The 2021 Strategy identifies four areas where additional action is needed to move the European economy towards sustainability: 1) Support the real economy transition by recognising and encouraging intermediary steps companies in certain economy sectors, especially energy and the gas industry, can take to align with EU environmental goals (transition finance); 2) Create a more transparent and credible framework for labels and financial instruments to prevent greenwashing (resilience and contribution); 3) Empower retail investors and SMEs to access sustainable finance opportunities (inclusiveness); 4) Foster global ambition by inviting international partners to deepen global cooperation on sustainable finance bilaterally and multilaterally, in particular to promote convergence of approaches and to provide the private sector with usable tools and metrics such as taxonomies (global consensus).

To execute its renewed sustainable finance strategy, the EU has put forward a package of legislative and regulatory measures, implementing bodies and further actions to mobilise the resources necessary to meet the European Green Deal investment needs.

Legislative and Regulatory Measures

The new sustainable finance framework introduces a mandatory disclosure regime consisting of the Sustainable Finance Disclosure Regulation 2019/2088 and its Implementing and Delegated Acts on sustainability-related disclosures in the financial services sector, and the Corporate Sustainability Reporting Directive (CSRD) 2022/2464 which expands the scope of Directive 2014/95/EU regarding sustainability disclosures for certain public interest undertakings. Both instruments are designed to help investors, consumers, policy makers and stakeholders evaluate and compare with more accuracy and confidence the sustainability performance of large and medium sized companies including private banks, and thus encourage more socially responsible business models and practices.

Another building block is the Taxonomy Regulation 2020/852 with its Delegated Acts which provide guidance and benchmarks to the private sector to track progress towards the greening of their operations. The EU Taxonomy is a common classification of economic activities that qualify as substantially contributing to climate change mitigation and adaptation and cause no significant harm to other EU environmental objectives. It represents the most comprehensive and advanced standard globally thus far (Janse & Bradford, 2021; OECD, 2020). The Taxonomy is complemented by the Benchmark Regulation 2019/2089, amending Regulation 2016/1011 as regards EU climate transition benchmarks, EU Paris-aligned Benchmarks and sustainability-related disclosures for benchmarks. With the Benchmark Regulation, the EU laid down the groundwork to define minimum EU standards that increase transparency and prevent incoherent, incomprehensible and misleading information provided to investors about low-carbon investment portfolios and sustainability indices.

Sustainability reporting requirements apply to large companies, whether listed or not, meeting two of the following size criteria: more than 250 employees and/or a turnover of more than €40 million and/or €20 million in total assets as well as companies with securities listed on an EU-regulated market, irrespective of whether the issuer is established in the EU or a non-EU country. Listed SMEs, with the exception of micro-enterprises, are allowed to report according to simpler standards than those that will apply to large companies; non-listed SMEs are encouraged to use the same standards as listed SMEs on a voluntary basis. CSRD requirements also apply to non-EU undertakings with annual EU-generated revenues in excess of €150 million and which also have either a large or listed EU subsidiary or a significant EU branch generating €40 million in revenues. The respective subsidiary or branch will be responsible for publishing sustainability reports for these non-EU undertakings at a consolidated level from 2028 onwards.

CSRD covers a total of about 50,000 companies and 80 percent of GHG emissions. However, for financial years starting on or after 1 January 2024, CSRD will apply to companies that are already subject to the Non-Financial Reporting Directive, with the first report expected to be produced in 2025. Large companies that are not presently subject to the Non-Financial Reporting Directive will have to apply CSRD from financial years starting on or after 1 January 2025 and therefore report in 2026 on 2025 data. For financial years starting on or after 1 January 2026, CSRD will be rolled out to listed SMEs, albeit subject to an opt-out until 2028, with the report in 2027 being based on 2026 data. Companies will be required to comply with detailed sustainability reporting standards – the European Sustainability Reporting Standards – developed by the European Financial Reporting Advisory Group. Such standards specify the information that companies should disclose with regard to all reporting areas and sustainability matters, and ensure alignment with regards to existing disclosure obligations set out in the Sustainable Finance Disclosure Regulation.

CSRD provisions ensure coherence with the EU Taxonomy and the Sustainable Finance Disclosure Regulation which applies to financial services institutions, i.e. banks, insurance companies, pension funds and investment firms. Reporting standards developed under CSRD include indicators that correspond to those contained in the Sustainable Finance Disclosure Regulation. So, for example, under the first Delegated Act implementing the Taxonomy adopted on June 4, 2021, the manufacturing of rechargeable batteries, battery packs and accumulators that result in substantial GHG emission reductions in transport, stationary and off-grid energy storage and other industrial applications, and the recycling of end-of-life batteries are considered economic activities that substantially contribute to climate change mitigation. Therefore, the Commission classifies them as environmentally sustainable provided they do not significantly harm any of the other environmental objectives set by the Taxonomy and comply with minimum social safeguards. Under the second Delegated Act, companies covered by CSRD must disclose the proportion of their turnover, capital expenditure and operating expenditures that are derived or associated with environmentally sustainable economic activities as defined by the EU Taxonomy. The sellers of financial products covered by the Sustainable Finance Disclosure Regulation are required to make pre-contractual disclosures and to report on how and to what extent the investments underlying the financial product are in environmentally sustainable economic activities according to the Taxonomy. These regulations create a system where both financial and non-financial corporate actors must follow a common language and shared definitions based on objective criteria giving investors and stakeholders a sense of what percentage of a company's activities or what portion of an investment portfolio is or not Taxonomy aligned (European Commission, 2023e, p. 4).

Implementing Bodies and Further Actions

The new sustainable finance framework consisting of a standardised set of mandatory reporting requirements, a common classification or Taxonomy, sustainability benchmarks and green labels is being further implemented by several *ad hoc* bodies and initiatives to address possible market inefficiencies and regulatory gaps.

With regards to mandatory reporting requirements, CSRD includes a general EU-wide audit (assurance) requirement to help ensure that the reported sustainability information is accurate and reliable. Given the current capacity and technical ability of the market for audit services, CSRD requires only a more limited assurance for the time being while allowing member states to open their markets for independent assurance service providers firms can choose from other than the usual auditors on financial information. The approach of limiting the scope of assurance in the initial stage of implementation of the ESG reporting system reflects a certain amount of caution on the part of EU institutions given the need to allow a sufficient lead time for the formation of adequate expertise and the professionalisation of new competences in the financial and accounting fields. At the same time, it shows a clear and strong commitment to treat sustainability information in par with financial statements. Companies will have to prepare their financial statements and their management report in XHTML format in accordance with the European Single Electronic Format Regulation and to "tag" their reported sustainability information according to a digital categorisation system as and when specified in that Regulation. The digitalisation of disclosures is expected to lower companies' reporting costs and improve the way investors and others use and compare information.

Another concern of EU institutions is how to make the EU Taxonomy more widely used and facilitate access to transition finance. Transition finance can be understood as the financing of climate and environmental performance improvements to transition towards a sustainable economy at a pace that is compatible with the climate and environmental objectives of the EU (European Commission, 2023c, preamble paragraph 5). The Commission first established the EU Platform on Sustainable Finance in 2020 as an expert group to advise on the development of the sustainable finance framework and in particular of the EU Taxonomy. The Platform identified several classification gaps and recommended the following actions: for instance, the Taxonomy does not assess which activities must be phased out in order to reach transition targets, nor which ones incrementally improve a company's performance towards net zero albeit not contributing *substantially* to climate mitigation and being, therefore, Taxonomy aligned. Such an approach limits participation by some sectors, companies or other economic actors who may have a valid transition plan. The Platform proposed ways to expand and develop the Taxonomy further to provide greater sector coverage and recognition of interim transition steps (Platform, 2021). In partial response to these recommendations, the Commission adopted amendments to the Third Delegated Act establishing additional technical screening criteria for determining the conditions under which certain economic activities qualify as contributing substantially to climate change mitigation or climate change adaptation and for determining whether those activities cause no significant harm to any of the other environmental objectives (European Commission, 2023d).

With regards to investment benchmarks and labels, in June of 2023 the Commission proposed a new regulation on the transparency and integrity of ESG ratings activities in light of growing investors and stakeholders concerns about the lack of clear, standardised, publicly available and sufficiently independent methodologies credit rating agencies use to incorporate ESG factors in their risk assessment evaluations of financial and non-financial issuers. The ESG rating market is highly decentralised and unregulated, and has recently been further compromised by some high profile scandals (Sipiczki, 2022). Members states do not regulate the activities of ESG rating providers nor the conditions under which they deliver their products or services and are likely to adopt divergent measures and approaches which would have a direct negative impact and create obstacles to the proper functioning of the single market and be detrimental to the ESG rating market. Given the need to offer a uniform level of protection to investors in different member states and foster comparability between ratings, under the proposed regulation, ESG rating agencies have to be authorised by and operate under the supervision of the European Securities and Market Authority. The regulation sets organisational, governance and transparency requirements, including the disclosure of methodologies, models and key rating assumptions used in ESG rating activities to the public, to guarantee independence and avoid potential conflict of interests between ESG rating providers and their service subscribers (European Commission, 2023b).

Although the EU Taxonomy and the standards, labels and disclosure requirements based on it are primarily designed for financial market participants, they can also be applied by the public sector. In fact, coherence between public and private sector investments is nothing but desirable (European Commission, 2023a). The European Investment Bank (EIB) supports sustainable finance policy in three fundamental ways. First, the EIB provides finance for climate action projects on favourable terms such as lower interest rates and longer repayment periods than borrowers would have been able to obtain on the market. Second, the EIB assesses compliance with its environmental and social standards which form a contractual condition for receiving financing. For instance, the EIB estimates the amount of potential greenhouse gas emissions and uses a shadow carbon price to factor into the project cost the external negative effects of the emissions, effectively reducing the economic viability of projects with higher emissions and their chances of being financed with EIB funds. The EIB also assesses the resilience of projects to climate change, and may require projects to be modified in order to limit their climate related risks. Third, the EIB promotes sustainable finance by encouraging other public and private investors to adopt sustainable finance good practices. The EIB is updating its due diligence procedures and has updated its climate tracking methodology to take account of the EU Taxonomy criteria. However, further action is needed to ensure greater transparency and alignment of EIB investment activities towards the financing of climate and mitigation goals, especially in central and eastern Europe where it is needed the most (European Court of Auditors, 2021, pp. 34–37).

The Environmental Conscience in Sustainable Finance

The EU sustainable finance framework fits the environmental conscience paradigm. First, as shown in the chapter, we begin to observe alignment with both energy and environmental policies in the field of finance. Sustainable finance regulations are designed to force financial institutions and businesses to measure their economic activity against pre-established ESG criteria and benchmarks, and to communicate that information clearly, truthfully and effectively with stakeholders and investors. In doing so, companies are required to follow a unique set of standardised requirements and make disclosures according to the principle of double materiality. Adopting a double materiality perspective to disclosure means that a company must report both on the impacts of its activities on people and the environment and on how sustainability matters affect its financial performance. Even if companies are not required to be Taxonomy-aligned, they must assess their operations based on a common, science-based, classification system. For instance, under the newly adopted set of guidelines, companies must report on certain climate and social performance indicators regardless of their materiality. If the information is material, they must also report on one or more of the six thematic environmental areas (climate change, pollution, water and marine resources, biodiversity and ecosystems, resource use and circular economy) as well as on a number of social and governance factors. The sustainability information so disclosed, in turn, becomes crucial for businesses to be able to access dedicated finance in the form of green bonds, tax rebates, subsidies and other financial benefits including more liquidity available to them in capital markets to implement their transition plans, and for financial institutions to design investment products more accurately and in accordance with investors' growing demand for sustainability-oriented funds.

Second, together with the EU Taxonomy, the European Sustainability Reporting Standards which incorporate the double materiality methodology represent absolute innovations in the field of corporate reporting. Because of their breath, depth, technical and scientific quality, and stakeholder-based approach these regulations are considered the "gold standard" among green taxonomies and sustainability standards globally (Tian Tong Lee, 2024). Via the International Platform on Sustainable Finance, the EU is leading the effort to foster dialogue between policy makers, exchange best practices and harmonise sustainable finance rules to scale up the mobilisation of private capital towards sustainable investments globally.

Third, the Commission's expertise in regulating financial markets dates back to the Union's founding and the need to guarantee the free movement of capital within the single market. The 2008 Global Financial Crisis further incentivised the Commission to complete the project of creating an Economic and Monetary Union (European Commission, 2020; Braun et al., 2018). The main concern being that capital markets in Europe are not sufficiently integrated so as to protect the EU and especially the Eurozone from future crises (Kaehler &

Weber, 2023; Lannoo & Thomadakis, 2019). At the same time, sustainability has long been central to the Union project. With the adoption of the European Green Deal, the Commission recognised that the financial system must gradually adapt to support the sustainable functioning of the economy. Establishing common criteria at the EU level for determining when an economic activity qualifies as environmentally sustainable coupled with Union-wide standards for environmentally sustainable financial products and their appropriate disclosure to investors are essential steps to eliminate barriers to cross-border transactions and facilitate the shift of investment towards environmentally sustainable economic activities to close Europe's climate finance gap.

Fourth, most Europeans see climate change as a serious problem. According to the latest Eurobarometer survey, more than eight in ten think the EU should take action to increase renewable energy and support energy efficiency, and the majority believe that the EU should accelerate the transition to a green economy in view of the energy crisis triggered by Russia's invasion of Ukraine (European Commission, 2023a). EU citizens also support greater transparency and accountability of corporate and financial actors. Consumers and stakeholders groups have been at the forefront of many lobbying efforts to pressure both companies and institutional investors to disclose more information on their ESG practices. Several surveys show a sizeable growing interest in ESG funds on the part of retail investors (Dupre, Bayer & Santacruz, 2020).

Going forward, we can expect EU financial markets to become more integrated towards the realisation of a Capital Markets Union and in support of a green and just transition. Furthermore, the incentivisation of green assets and acceleration towards a sustainable economy would put Europe on a path towards greater energy independence vis-à-vis Russia. However, the current sustainable finance framework could be further strengthened. Substantial gaps and discrepancies exist in the Taxonomy (Schütze & Stede, 2021). In an effort to address some of those gaps, with the third Delegated Act the Commission included specific nuclear and gas energy activities in the list of economic activities classified as sustainable under strict conditions. Specifically with regards to electricity generation, co-generation of heat/cool and power and other heating/cooling systems powered by natural gas, screening criteria have so many qualifications that the average facility would not meet the threshold unless introducing substantial value chain methane leak detention and carbon capture and storage technology to reduce emissions. Unfortunately, methane leaks are wildly underreported by companies. The International Energy Agency (IEA) estimates that global energy sector methane emissions are seventy percent higher than official figures (IEA, 2022). When fully accounted for, the carbon intensity of gas-fired power nears that of coal. Unless those reporting gaps are addressed, continued utilisation of natural gas until 2035 as currently contemplated by the Taxonomy may raise both investment and decarbonisation risks. Including those activities in the Taxonomy may also reduce the amount of spending in greener, more worthy, projects. Additionally, the energy sector remains dominated by fossil fuel spending, and oil and gas companies have little trouble accessing capital to finance their operations in and outside the EU. The Taxonomy should, therefore, explicitly exclude new unabated fossil fuel infrastructure or investments in accordance with the scientific findings of the Intergovernmental Panel on Climate Change and IEA. Instead, EU sustainable finance policy is silent with respect to the financing of new coal and oil production. While companies can rely on the Taxonomy classification when reporting on their sustainability performance and as a tool to plan their climate and environmental transition; and financial market participants can use it to design green financial products; "there is no obligation for companies to be Taxonomy-aligned and investors are free to choose what to invest in" (European Commission, n.d., p. 1). The Commission insists that those regulatory actions will have "indirect positive effects on respect for fundamental rights, as well as on people and the environment, since more stringent reporting requirements can influence corporate behavior for the better" (European Commission, 2021b, pp. 10, 12). The reality is that additional complementary transition policies are going to be needed to sufficiently incentivise companies to decarbonise. This means not only requiring that all economic activities align with decarbonisation scenarios but also phasing out subsidies for brown assets, setting sunset clauses for existing fossil fuel power plants for which alternative technologies exists, establishing clear criteria on how to deal with stranded assets and higher caps on carbon pollution. One step in this direction is the Corporate Sustainability Due Diligence Directive pending publication in the Official Journal of the EU (European Commission, 2022). The Directive sets out an obligation for large companies to adopt and put into effect, through their best efforts, a transition plan for climate change mitigation aligned with the 2050 climate neutrality objective of the Paris Agreement as well as intermediate targets under the European Climate Law (European Commission, 2022, pp. 4–5). The Directive's aim is to complement corporate sustainability reporting requirements by adding a substantive duty to identify, on the basis of information reasonably available to the company, the extent to which climate change is a risk for, or an impact of, the company's operations, and if so, include emissions reduction objectives in its plan (article 15, paragraphs 1 and 2).

Another area for improvement is public procurement. The Public Procurement Directive (2014/24/EU) permits the inclusion of environmental considerations at various stages of the public procurement procedure but does not set binding requirements. Because of the lack of comprehensive mandatory targets at the EU level, the amount of green public procurement taking place in the Union largely depends on decisions by the member states and their public bodies which vary widely from country to country (Sapir et al., 2022). The role of public procurement in boosting innovation, creating lead markets and ultimately change production patterns is well established in economic analysis. Thus far, EU institutions have chosen a limited sectoral approach (e.g. Clean Vehicles Directive 2019/1161/EU; Energy Performance of Buildings Directive 2010/31/EC; Energy Efficiency Directive 2023/1791/EU). Adopting binding green procurement requirements including a specific requirement for public pension funds and other large public investors to work toward Taxonomy-aligned investment portfolios would add another critical piece to the 2050 carbon neutrality puzzle.

Conclusion

EU sustainable finance policies and actions are part of a more general European environmental conscience which has been evolving over several decades as the EU implemented various policies and regulations to protect the environment and promote sustainability. A new area of EU regulation, sustainable finance serves the dual purpose of supporting energy and environmental goals at home while also advancing EU leadership in climate change governance on the global stage. Financial regulation at the EU level takes on a new urgency in the context of building a more sustainable economy because more developed and integrated capital markets are essential to unlock investments for companies, especially SMEs, and provide funding for innovative projects. Given the immediate priority to make progress on climate mitigation and raise the resources necessary to transition to a low-carbon economy, the focus of the current sustainable finance policy framework is on closing the EU climate finance gap, including by providing capital to those economic activities that while not contributing substantially to abate emissions are deemed to advance the transition at a pace compatible with EU climate and environmental objectives. Such a broad definition of what constitutes a *sustainable economic activity*, coupled with no requirement on the part of economic and financial actors to align with Paris Agreement benchmarks, risks compromising the attainment of climate goals. A related risk is reducing sustainability to a mere financial opportunity rather than a means for social transformation. Via the monitoring and strengthening of the EU Taxonomy and other parts of the sustainable finance framework, EU institutions are committed to make progress towards a more holistic conceptualisation of sustainability in finance. Whether EU sustainable finance policy will result in the financialisation of sustainability rather than true sustainability in finance remains to be seen.

Bibliography

Ahlström, H., & Monciardini, D. (2022). The regulatory dynamics of sustainable finance: Paradoxical success and limitations of EU reforms. *Journal of Business Ethics*, *177*(1), 193–212.

Braun, B., Gabor, D., & Hübner, M. (2018). Governing through financial markets: Towards a critical political economy of Capital Markets Union. *Competition & Change*, 22(2), 101–116.

Braun, T. (2024). Financial Enforcement Instruments of Sustainability. In M. Tendero & C. Weber (eds), *The European Environmental Conscience in EU Economics: Finance, Innovations, and External Relations*. Routledge.

De Schutter, O. (2008). Corporate Social Responsibility European Style. *European Law Journal* 14(2), 203–236.

Dupre, S., Bayer, C. & Santacruz, T. (2020). A large majority of retail clients want to invest sustainably. *2° Investing Initiative*. https://2degrees-investing.org/resource/retail-clients-sustainable-investment/

European Central Bank. (2022). Results of the 2022 climate risk stress test of the Eurosystem balance sheet. https://www.ecb.europa.eu/pub/economic-bulletin/focus/2023/html/ecb.ebbox202302_06~0e721fa2e8.en.html

European Commission. (2011). A renewed EU strategy 2011–14 for corporate social responsibility. COM (2011) 681 final. 25 October 2011.

European Commission. (2014). Disclosure of non-financial information: Europe's largest companies to be more transparent on social and environmental issues. STATEMENT/14/124/. 15 April 2014.

European Commission. (2015). The five presidents' report: Completing Europe's economic and monetary union. https://commission.europa.eu/publications/five-presidents-report-completing-europes-economic-and-monetary-union_en

European Commission. (2018). Action plan: Financing sustainable growth. COM (2018) 97 final. 8 March 2018.

European Commission. (2019). The European green deal. COM (2019) 640 final. 11 December 2019.

European Commission. (2020). A Capital Markets Union for people and businesses-new action plan. COM (2020) 590 final. 24 September 2020.

European Commission. (2021a). Strategy for financing the transition to a sustainable economy. COM (2021) 390 final. 6 July 2021.

European Commission. (2021b). Proposal for a directive as regards corporate sustainability reporting. COM (2021) 189 final. 21 April 2021.

European Commission. (2022). Proposal for A Directive On Corporate Sustainability Due Diligence And Amending Directive (EU) 2019/1937. COM (2022) 71 final. 22 February 2022.

European Commission. (2023a). Citizen support for climate action. https://climate.ec.europa.eu/citizens/citizen-support-climate-action_en

European Commission. (2023b). Proposal for a regulation on the transparency and integrity of environmental. Social and Governance (ESG) rating activities. COM (2023) 314 final. 13 June 2023.

European Commission. (2023c). Commission recommendation on facilitating finance for the transition to a sustainable economy. 27 June 2023.

European Commission. (2023d). Delegated regulation establishing additional technical screening criteria for determining the conditions under which certain economic activities qualify as contributing substantially to climate change mitigation or climate change adaptation and for determining whether those activities cause no significant harm to any of the other environmental objectives. 27 June 2023.

European Commission. (2023e). Enhancing the usability of the EU taxonomy and the overall EU sustainable finance framework. SWD (2023) 209 final. 13 June 2023.

European Commission. (2024). Report on the macroprudential review for credit institutions, the systemic risks relating to non-bank financial intermediaries and their interconnectedness with credit institutions. COM (2024) 21 final. 24 January 2024.

European Commission. (n.d.). FAQ: What is the EU taxonomy and how will it work in practice? https://finance.ec.europa.eu/system/files/2021-04/sustainable-finance-taxonomy-faq_en.pdf

European Council. (2019). Council conclusions on the deepening of the capital markets union. 14815/19. 5 December 2019.

European Court of Auditors. (2021). Sustainable finance: More consistent EU action needed to redirect finance towards sustainable investment. https://www.eca.europa.eu/en/publications?did=59378

European Parliament. (2020). Corporate social responsibility (CSR) and its implementation into EU company law. https://www.europarl.europa.eu/thinktank/en/document/IPOL_STU(2020)65854

European Union. (2013). *Accounting directive*. https://eur-lex.europa.eu/eli/dir/2013/34/oj

European Union. (2014). *Non-financial reporting directive*. https://eur-lex.europa.eu/eli/dir/2014/95/oj

European Union. (2019). *Sustainable finance disclosure regulation*. https://eur-lex.europa.eu/legal-content/EN/TXT/?uri=CELEX:32019R2088

European Union. (2020). *Taxonomy regulation*. https://eur-lex.europa.eu/eli/reg/2020/852/oj

European Union. (2022). *Corporate sustainability reporting directive*. https://eur-lex.europa.eu/legal-content/EN/TXT/?uri=CELEX%3A32022L2464

Hoerber, T. (2012). *The origins of energy and environmental policy in Europe: The beginnings of a European environmental conscience*. Routledge.

Hoerber, T., & Weber, G. (2022). *The European environmental conscience in EU politics: A developing ideology*. Routledge.

International Energy Agency. (2022). *Global methane tracker 2022*. https://www.iea.org/reports/global-methane-tracker-2022

Janse, K. A., & Bradford, A. (2021). The Brussels effect on sustainable finance. *Project Syndicate*. https://www.project-syndicate.org/commentary/eu-sustainable-finance-taxonomy-brussels-effect-by-kalin-anev-janse-and-anu-bradford-2021-04

Kaehler, J., & Weber, C. S. (2023). European monetary policy in a globalized world: From low to high politics. In *The EU in a globalized world* (pp. 115–131). Routledge.

Lannoo, K., & Thomadakis, A. (2019). Rebranding capital markets union: A market finance action plan. CEPS-ECMI Task Force Report, Centre for European Policy Studies.

Nadakavukaren-Schefer, K. (2020). Social power, social responsibility and corporations: From CSR to business and human rights. *ZSR, 139*(II), 5–95.

OECD. (2020). Developing sustainable finance definitions and green taxonomies. https://www.oecd.org/env/developing-sustainable-finance-definitions-and-taxonomies-134a2dbe-en.htm

Platform on Sustainable Finance. (2021). Transition finance report. https://finance.ec.europa.eu/document/download/c5e91dc2-7a28-4a30-aae9-9fd667195d28_en

Ruggie, J. G. (2011). Guiding principles on business and human rights: Implementing the united nations "Protect, Respect and Remedy" framework. Report n. A/HRC/17/31. United Nations.

Sapir, A., Schraepen, T., & Tagliapietra, S. (2022). Green public procurement: a neglected tool in the European green deal toolbox? *Intereconomics, 57*(3), 175–178.

Schütze, F., & Stede, J. (2021). The EU sustainable finance taxonomy and its contribution to climate neutrality. *Journal of Sustainable Finance & Investment, 14*(1), 128–160.

Sipiczki, A. (2022). A critical look at the ESG market. *CEPS Policy Insights*. No 2022-15/April 2022.

Sparkes, R. (2006). A historical perspective on the growth of socially responsible investment. In Rory Sullivan and Craig Mackenzie (Eds.), *Responsible investment*. Routledge.

Sparkes, R., & Cowton, C.J. (2004). The maturing of socially responsible investment: A review of the developing link with corporate social responsibility. *Journal of Business Ethics 52*, 45–57.

Tendero, M. (2021). The time has come! The development of the European environmental conscience. Evidence from the Eurobarometer surveys from 1974 to 2020. In T. Hoerber, & G. Weber (Eds.), *The European environmental conscience in EU politics* (1st ed., pp. 185–206). Routledge.

Tendero, M. & Weber, C. S. (2024). *The European environmental conscience in EU economics: Finance, innovations, and external relations*. Routledge.

Tian Tong Lee, S. (2024). *China proposes new ESG rules to keep up with Europe*. Bloomberg.

Weber, C. S. (2015). La Euro crisis: causas y síntomas. *Estudios fronterizos*, 16(32), 150–172.

4 Financial Enforcement Instruments of Sustainability

Tomasz Braun

Introduction

There are numerous regulatory tools of multiple character that can be used as means to enforce sustainability policies that lead to expected sustainable businesses conducts. Many multilevel financial regulatory tools that are used to assure sustainability enforcement can take various forms as soft discouragement or hard incentives (Weiss & Kammel, 2015). They range from such incentives as tax levies, subsidies and grants, preferentially structured green bonds, CO_2- and renewable energy–related financial instruments, environment protecting covenants in financing agreements up to non-financial (sustainability) publicly disclosed reporting obligations. They can have different reach and varying efficiency (O'Farrell & Anderson, 2010, 62). All of them feed into the universe of necessary actions that help the implementation of sustainability. Indeed, the implementation and thereafter the enforcement of policies being the greatest challenge make the regulators seeking different ways to assure them. Among them there is a number of sustainability implementation takeaways assumed from the global corporates that can be useful in the process of regulatory policies enforcement (Rishi, 2022, 150). They can be as simple as the government (or governmentally sponsored) advisory targeted to wider public, democratically engaging semi-innovative participatory instruments that precede hard regulating, they can also be other instruments of regulatory control like sandboxing, auditing, and enforcing through various nonobvious and indirect (i.a. political and financial) measures.

Many of these measures have already been tested while regulating financial industry. Multicentricity of regulations as a remedy to inward-looking policies is one of them (Benz, 2021, 148; Zeitlin, 2023, 84). The more stakeholders such as civil society organisations, businesses, financial institutions, media etc. become part of the sustainability enforcement the more there are chances for their efficient implementation (Quaglia, 2020, 51).

This text is another voice in the discussion within the sustainability debate. It is also intended to provide a reference for these who realise that contemporary governance instrument need to be introduced and enforced on multiple levels (Vantagiatto, 2020, 183). The opinions presented here are hoped to serve

DOI: 10.4324/9781032656359-7

as a support in a search for efficiency of regulatory tools in a normative legal theory discourse. This is particularly timely in the context of a broader discussion of the European Environmental Conscience being the guiding theory for this book (Weber, 2025 in Chapter…of this book).

The research has been conducted from two perspectives: multilevel governance analysed from the normative theory viewpoint and a potential of financial instruments to be used within the sustainability enforcement. It combines both areas and illustrates interrelations among them. It also proves that the major global challenge, as the sustainability enforcement is, requires novelty of the regulatory approach which can benefit from the practices tested in the financial institutions' environment.

The chapter has been divided into four parts. The first focuses on analyses of existing multi-layered financial enforcement measures. It refers to the fact that use of multileveled financial instruments that used to be innovative some time ago becomes a regulatory reality of in various spheres. At present is increasingly deployed in relation to selected specific environmental agendas like decarbonisation, energy transformation, circularity of resources, deforestation, wildlife and water protection. Then it discussed to what extend the proposed approach boosts enforcement tools that have been successfully tested earlier in other areas.

The second part describes sustainability policies enforcement takeaways from financial institutions. It proves that involvement of engaged actors, specifically businesses and financial institutions is instrumental as there is a particular regulating role to be played by them. In particular within their internal policies and practices. Thereafter their disseminating impact and promoting effect among the addressees cannot be neglected. The section illustrates the problem of the EU policies enforcement from the perspective of the lessons driven from the financial institutions as public trustees that fulfil their role of the enforcement actors mandated by the public authorities.

The third part relates to interdependencies of corporate practices and public policies on sustainability enforcement. It focuses on the problem to what extend privatised regulations are connected to public ones and whether lawmakers choose from among various available non-public regulatory instruments. The enrichment of hard laws by the soft law instruments in the process of environmental enforcement has proven to be useful in countering numerous irregularities. It analyses if enforcement efficiency is enhanced once multijurisdictional corporates work together with public authorities to create a system for implementation of sustainable laws, regulations and policies.

The fourth part of the chapter reflects on a need for unification of sustainability enforcement tools. As interrelations of different types of regulation and multileveled approach supports efficiency of the laws, regulations, and policies it might be considerable to seek for a methodical approach to consolidating them. First examples of this approach have been identified within the EU initiatives. The concluding postulates of the chapter direct to closing remarks that important part of the EU sustainability agenda plays an economy boosting role and that especially the use of financial tools deserves further thorough exploration.

Multi-layered Financial Enforcement Measures

Usually, global sustainability enforcement problem is addressed either from purely political or from the socio-economical angle. It is less obvious to discuss it from the point of view of multilevel regulatory governance and financial tools. Nowadays, using innovative multileveled financial instruments becomes a regulatory reality of in various spheres. What happens in sustainability regulations now will have a significant impact on the quality of live, culture and economic situation of the societies in future.

Analysis of financial sustainability enforcement measures should take into account the fact that it is a dynamic and multilevel process. It changes in time and manifests in various forms depending on which enforcement tools are being used. There is a growing number of sustainability enforcement players involved in this process. In the result, there are varying interests and multi-vectoral dynamics among them. Numerous stakeholders engaged in sustainability enforcement are most evident on the European plan. The debates among the EU and its Member States as well as among Member States and other players active on the single market, other organised societies groups within the EU, the external EU relations with the third countries that more or less willingly partner in these debates – all of them take place simultaneously and by definition are of neuronal nature.

The multilevel enforcement measures debates expand also outside of the EU as the unsustainability problems are global. As much as a need for universal measures is not questionable, the search for globally efficient tools have been also addressed by the United Nations and its agencies. In fact, the UN was first and other organisations as the Council of Europe and the EU followed. It is little debatable that given the limited power of the United Nations, only when they have been cascaded down to the EU and then the Member States' levels, the regulated measures could have been deployed more efficiently (Hoerber et al., 2021, 51). Still their efficiency remains uneven. The fact regards in particular the present cross-border sustainability policies' enforcement that refer to environmental issues discussed in the EU.

There are numerous stakeholders engaged in sustainability discussions. Among them there are intergovernmental organisations, their members and their markets, the states and the societies, the press, other organisations, third countries and the organised activists (Alexander, 2022, 4). As a result, a number of the internationally impactful environmental policies have been adapted at multiple levels (Fernandez, 2021, 94). They are usually theme-focused on selected pro-climate agendas like emissions' reduction, renewable energy, waste management, deforestation, protection of waters etc (Grolleau et al., 2022, 107307). Various political agencies present their positions to assure their implementation.

A wide acceptance of sustainability goals is an unprecedented opportunity. However, for a worldwide enforcement of sustainability policies, the challenges on the way of this process need to be considered. They are threefold and refer to the language of norms, proposed regulatory methodology and the selection of enforcement instruments. Different understanding of basic notions, and uneven implementation of them, due to varying local specificities and interests may conclude in various levels of tightness of national enforcement systems. Therefore, global sustainability enforcement risks to be an unresolved problem. To make it more efficient it is necessary to use more universal measures. The proposed approach suggests the deployment of enforcement tools that have been successfully tested earlier in other areas.

The choice of these tools should take into account a general need of multi-leveled approach to policymaking and governance, as well as the impact of global financial institutions on societies and the potential of using them as agents in the process of enforcement. Combining both elements and having considered interrelations among them means a novelty of the regulatory approach which gives a quality boost in the enforcement process. To do so lawmakers need to learn from the corporates and also from other organised societies' experience (Barnes & Hoerber, 2013, 7). The search for efficiency improvement and a discussion how the normative legal theory can enhance this improvement through for the benefit of regulatory enforcement is an opportunity to take. In this context it needs to be mentioned the general jurisprudence debate around the efficiency of lawmaking methodologies, namely if the laws and regulations should be granularly casuistic and thus encourage some and discourage other activities or if they rather stay generic and let the addressees of the norms liberty to select how they want to achieve the set objectives (Lehman, 2020, 75).

Multiple governance tools that proven their efficiency are based on the principle of initial consultation, participation and engagement followed by controls and post factum monitoring of its implementation. Similarly, multilevel enforcement can be effectuated throughout a variety of measures as audits and checks that result in recommendations assured by executable fines. These measures have already been tested while regulating irregularities and wrongdoings of financial institutions.

From the enforcement assurance perspective, it is a relatively new process aimed at providing a well-thought, efficient and just regulatory framework to address current sustainability challenges. Within this process of rethinking the enforcement tools the focus should be put on financial institutions and more generally on other global corporates. The overarching presumption is that if takeaways of global corporate policies have proven efficiency then the same may refer to regulations responding to problems caused by unsustainable development of industries that turned to be detrimental to environmental, economic and societal spheres. Therefore, the efficiency of enforcement being at stake, much can be learned from the corporate world as for example the approach to compliance norm-making and implementation practices referring to them. They can serve as repeatable instruments and tested benchmarks to sustainability enforcement (Heddeman-Robinson, 2018, 42).

In contrary to that, the inefficiency of inward-looking policies and underuse of the public engagement makes the enforcement less efficient than it could be if more creative approach has been deployed. It is crucial therefore to engage financial institutions, businesses, and civil society organisations into implementation and enforcement of the sustainability goals. The policymaking deploying the multilevel enforcement model means engagement of various stakeholders and use of multiple tools of a cross-border impact to enforce the policies on the supranational level. A particular focus should be put on the instruments at hands of the financial institutions and more generally on the power possessed by the global corporations that in many instances proved to be even more successful in achieving their goals than public authorities were (Pereira et al., 2021).

The well-known concepts of ubiquitous regulatory tools come back to the discussion nowadays. The soft norms implemented by various stakeholders at different levels can lead to favourable enforcement results but require an openness from the policymakers (Knodt, 2019, 181). The task refers to a widely known concept of compliance implemented by corporates long time ago. It relates to the role of corporate governance, internal conducts codes and the risk policies introduced the normative environment by the global corporations (Sheehy, 2019, 80; Braun, 2019, 123).

Sustainability Policies Enforcement Takeaways from Financial Institutions

An innovative approach to sustainability enforcement relates to the use of financial instruments. For their efficiency involvement of engaged partners, including businesses and financial institutions is needed. Their role is instrumental in relating governance policies to corporate practices and then in promoting them widely among the addressees of these policies.

The same relates to prudential requirements put on financial institutions as a result of hard lessons from financial crises that can serve as reference for sustainability enforcement. Implementing them required public acceptance leading to introducing relevant laws and regulations. They were preceded by engaged and informed debates in which public actors played a pivotal role as they are the ones able to implement financial enforcement tools (Colgan et al., 2021, 588). The same may refer to other market-related instruments that support sustainability enforcement. They can be directed to those referring for financial support from the financial institutions.

Among the financial measures that can have an enforcement power are the public debt instruments. The innovations of sustainable finance developed recently are green, sustainability and social bonds. Although they are constructed similarly to other types of investment bonds the difference is in the purpose of their use (Hill, 2020, 27). They are issued to finance sustainability impactful investments and cannot be used otherwise (Sachs et al., 2020, 7). They can only be used for financing or re-financing projects, assets or products that are in line with their purpose and not solely to satisfy return on investment expectations. Their specificity is also that the buyer has a recourse to the balance sheet of the issuer in case the bonds have not been used in accordance with their purpose and at the same time. Additionally, they allow to mitigate the financial risks of the sustainability projects financed through labelled bond finances (Maltais & Nykwist, 2020).

The challenge the sustainably bonds respond to is a limited access to long-term financing for investment in impact projects (Glavas & Okala Onana, 2025 in Chapter…of this book). The sustainability bonds can be issued by private and public financial institutions. To tackle greenwashing and to set up comprehensible standards for issuers and investors, the EU introduced European Green Bond Standard Regulation meant to assure assets needed for the low-carbon transition. This voluntary standard relies on the detailed criteria of the EU Taxonomy Regulation ((EU) 2023/2631 of the European Parliament and of the Council of 22 November 2023 on European Green Bonds and optional disclosures for bonds marketed as environmentally sustainable and for sustainability-linked bonds). As much as the European Green Bond Standard relates to all types of institutions, the mission of public financial institutions makes them better accommodated to respond to these limitations and to support the sustainability projects. The banks and other intergovernmental financial institutions as the European Bank for Reconstruction and Development and the European Investment Bank provide hundreds of billions of euros of available financing for these purposes. They provide equity and debt instruments covering costs of transition of economies by injecting liquidity and covering cost of risk guarantees for energy efficiency transformations, renewable energy investments, and sustainable transport projects (Sugiawan & Manag, 2019, 9).

There are more financial schemes by sustainably engaged financial institutions that open dedicated business lines for such products and projects. Additionally, they integrate sustainability considerations into conventional project financing and build into them preferential pricing mechanisms. They also use sustainability criteria, measurement, monitoring, and auditing to evaluate performance indicators such as greenhouse gas emission reductions (Ziolo et al., 2021, 38). In parallel to standard approval procedures, they supplement them with more sophisticated methodologies referring to sustainability risk management. All these mechanisms differ depending on specific industry sectors and supported technology. They vary but in the first instance they refer to renewable energy production, storage, transport, waste management, wildlife preservation, water protection, climate related innovations, circular economy etc. The support refers to project on different levels of implementation from research and development to deployment and modernisation (Taghizadeh-Hesary & Yoshino, 2020, 788).

Engaging financial institutions in sustainability implementation is built on a concept of delegating enforcement one level down from public institutions. It may be regarded as a *sui generis* subsidiarisation. Sometimes it is also defined as privatisation of law enforcement. It is usually effectuated through formalising contractual obligations by introducing sustainability covenants in investment, trade, or project finance agreements. Another instance of deployment of financial institutions as intermediaries in sustainability enforcement process is also executed in the fit and proper testing of risks management. In this context, the contractual mandates that financial institutions have over their counterparties in reviewing the execution of agreements is a method of extended sustainability enforcement.

Other financial instruments that have become sustainability enforcement measures have proven to be even more efficient due to their globally tradable character. The EU Emissions Trading System (ETS) introduced in 2005 instruments like green and renewable energy certificates are the examples of them. Established as s a result of public regulatory policies supporting the Clean Development Policy (CDM) with the European Commission (EC) adoption of the legislative measures to attain climate neutrality (Eyl-Mazzega & Mathieu, 2019, p. 25). They have been phased and planed for achieving the net 55% greenhouse gas emissions reduction by 2030 (Busch et al., 2021, 15). The pioneering character of CDMs pathed the way to engage financial institutions to a greater play of sustainability enforcement. Vastly inefficient, it mobilised regulators to replace it with a tighter mechanism. As a result, the Paris Agreement established a new market mechanism that replaced the CDM system and Joint Implementation mechanism after 2020. In the interim period, subject to some restrictions, participants in the ETS could use international credits from CDM and JI as part of their obligations under the upcoming EU ETS.

Among other changes that the EU regulations introduced there are the schemes to reach the EU climate targets under the European Green Deal that include the ETS, effort sharing targets within the Greenhouse Gas Emission Reductions Regulation,[1] EU rules on land use, land use change and forestry (LULUCF),[2] Sustainable Finance Disclosure Regulation (SFDR),[3] and also other standards and classification systems like EU Labels for Benchmarks (climate, ESG) and Benchmarks' ESG Disclosures,[4] EU taxonomy for sustainable activities,[5] European Green Bond Standard Regulation[6] etc. Particular importance and a great change in sustainability enforcement is expected to be achieved by introducing EU Corporate Sustainability Reporting Directive (CSRD) that entered into force on 5 January 2023.[7] To comply with it certain undertaking will have to disclose specific information about the following environmental factors: climate change mitigation, including the scope of reduction of greenhouse emissions; climate change adaptation; water and marine resources; resource use and circular economy; pollution; biodiversity and ecosystems.

The role of technology in achieving sustainability goals continues to be even more instrumental in the context of the support of registering and clearance of the offset mechanisms. Much of these could happen with the support of fintechs and other innovative financial institutions that continuously are being engaged in the overall sustainability enforcement process (Macchiavello & Siri, 2020, 6). As a result, they are growingly used in the analytical processes that use big data and artificial intelligence to assure more accurate and reliable calculation of impacts and offsets.

Sustainability enforcement through financial institutions refers to more than offsets and accuracy of data that assures reliability of clearance. Additional role is played by subsidies and other types of direct financial support provided to industry or individuals by lowering the price of certain technologies, energy, waste or promoting sustainability favourable behaviours. Soft subsidies may also take the form of research and development programmes. The downside of the direct subsidies is that although they are effective and their results are immediately visible, they come at the expense of taxpayers. A critical question therefore refers to the level of financial support as the amounts must encourage switching to new technologies or sustainable behaviours, but their public costs cannot be ignored. As a result, as much as the subsidies are efficient the burden of the financing is distributed among taxpayers and not put on the polluters (Tanaka, 2011, 6547).

Fiscal policy as imposition of taxes, tax rebates or tax exemptions and subsidies can impact the enforcement of sustainability. In parallel, fuels subsidising has been a popular policy instrument of a dual use: it increases affordability of and access to energy on one hand and can help shaping the energy sources mix in the overall energy balance. This relates in particular to gas and gas-related fuels. At the same time most of these instruments endanger competitiveness. This proves to be more effective if they are combined with direct tax exemptions. Taxation strengthens the efficiency of sustainability enforcement and limits emissions, but it may also endanger free competition. It is a dilemma that public authorities have to address to assure the fairness and equity of the rules (Sarker et al., 2020, 6). Energy efficiency and renewable energy are essential measures to control CO_2 emissions (Grolleau & Weber, 2024). Fiscal policy as well as other CO_2 and renewable energy-related financial instruments are vital for energy generation, storage, and consumption. The relationship between energy efficiency and CO_2 emissions within fiscal policies is clear. Fiscal policy that promotes renewable energy installations deteriorates the CO_2 emissions, but it can impact the economic growth which by itself may not be as bad as the political risks associated with this.

Both the fiscal and subsidy measures such as InvestEU, Innovation Fund, REPowerEU, or the Strategic Technologies for Europe Platform can be efficient once preceded by the stakeholders engaging informed participatory debate. An example of that a debate is enforcement of Trade and Sustainable Development (TSD) chapters within the EU Free Trade Agreement that have been drafted upon a consensus on what steps would lead to the goals set by the document. There has been a list of practicable actions and recommendations falling into the categories of working together; enabling and civil society including the social partners to play a greater role in implementation; delivering; and transparency and communication. A subsidies rulebook to be worked out by the EC and the World Trade Organisation to facilitate the fair access to financial support by foreign investors is an important factor of that communication.

An innovation proposed in the TSD chapters ultimately aim at civil society engagement to respond to the climate change challenges through the debate on improvements of sustainability enforcement. The EC have been mandated to review the impact of the implementation of TSD chapters together with other stakeholders and is expected to engage to analyse the effectiveness of the TSD chapters enforcement.

Apart from a number of fiscal policy sustainability enforcement measures there are the environment protecting covenants in financing agreements. It means increase of costs non-sustainability and therefore incentivising borrowers to adhere to sustainability regulations. Sustainability covenants trigger actions to avoid e.g. environmental damages or take remedial actions when these damages occur. They allow lenders to require necessary information and conduct audits to evaluate sustainability risks. In case there are events of breaching of environmental covenants, they have right to impose more rigid and expensive loan terms. They may also freeze or cancel financing. If the borrowers comply with sustainability regulations, they avoid these consequences (Akbar et al., 2021, 2605).

Interdependencies of Corporate Practices and Public Policies Sustainability Enforcement

Sustainability is only as much implemented as the obligations to do so are translated into the laws and even the best written laws are only as effective as they are enforceable. It is for lawmakers and regulators to use appropriate measures to assure the laws and regulations are complied with. The global dependencies continue to complicate the landscape among the jurisdictions. The size of non-sustainability damages caused by actors from across and despite of the geographies continues to grow as much as grows the societies' awareness of it (Ganguly et al., 2018, 852). The liability question for adverse effects of actions or negligence vis-à-vis the sustainability remains open despite of the fact that harmful effects of unsustainability across uncounted number of industries, activities, geographies, and actors keep increasing.

To address it rightly requires involvement of multiple stakeholders. Involving them prior to issuing regulation is the best lawmakers and regulators can do to assure participatory engagement (Bellucci & Manetti, 2018, 111). This is not just a right thing to do from the democratic dialogue perspective (Fiorino, 2018, 92). It can also help assuring the regulations are a better fit for purpose and it increases the probability these regulations would be more complied with. It is therefore for lawmakers and regulators to select from a variety of instruments to correctly adjust them. Apart from hard law, this selection would refer to the soft law instruments as they may be more efficient than other well-known normative tools. This approach is being re-invented by lawmakers and regulators who observe and replicate corporates' practices. Compliance as a concept referring to the way the addressees of the norms adhere to them has been known since long and continues to be used (Rogers, 2022, 75). It refers more widely than just to the large international corporations. The unobvious legal and regulatory instruments that boost the enforcement process could and should be used more extensively.

Introducing a new and more flexible approach to enforcement, also in a case of implementing corporate policies tools, needs a broader debate. Importance and relatively wide acceptance of sustainability allows for the use of multilevel enforcement tools as well expertise and experience in implementing these laws and regulations. One of many takeaways from the corporates' is that sustainability enforcement requires coordinated cooperation of various actors globally. Uneven and relatively weak enforcement mechanisms of sustainability laws and regulations are facts even within such a unified organisation like the EU (Rossinioli et al., 2021, 32). The EU laws and regulations refer equally to all subjects operating within its territory, but the efficiency of enforcement processes varies among the Member States. It is a missed opportunity given the fact that properly implemented multilevel enforcement of sustainability programmes could play an integrating and economy boosting role of the EU. The use of financial enforcement tools in this process deserves to be more thoroughly explored.

There are well tested methods which measure, control, assure and correct prudent approach of businesses to selected matters and its compliance with regulatory norms. An example is the use of fit and proper testing that can be used for addressing sustainability-related risks within the process of policies implementation (Bastedas-Orteaga & Stewart, 2019, 20). Another example is the obligation to introduce rigorous approach to sustainability measures and enforcement indicators together with accountability of people responsible for this (Waas et al., 2014, 5530). It has been imposed on financial institutions in reference to their stability indicators after having discovered irregularities in their reporting control mechanisms during the financial crisis of 2008 and it has proven to be working since then. Similar mechanisms can be applied to sustainability enforcement based on the principle of compliance with relevant obligations that have to be measured and controlled through application of measurable indicators. This process can easily be supported by growingly powerful technology solutions. An increasing role of sustainable (impact) investments in fintech and the sustainable financial institutions sector has been a boosting element of this process (Omarova, 2020, 78; Machiavello & Siri, 2022, 130). Another corporate practice of sustainability enforcement is disclosure of sustainability policies in prospectus regulations and within various publicly disseminated corporate reports and other documents (Gargantini & Di Noia, 2020, 370). The management report should consist of information necessary to understand the undertaking's impacts on sustainability matters, and information necessary to understand how sustainability matters affect the undertaking's development, performance, and position. This information related to sustainability matters should contain inter alia a business model and strategy, a description of the time-bound targets, a description of the role of the administrative, management and supervisory bodies, a description of the undertaking's policies, information about the existence of incentive schemes, a description of the principal risks and indicators relevant to the disclosures. In these cases, accountability for delivery of sustainability indicators is the condition under which the disclosed corporate reports have been useful as sustainability enforcement tools (Lepore & Pisano, 2022, 28).

The lessons learned from the corporates' practices have proven to be useful while countering any times of wrongdoings and misreporting (Ghoogossian, 2015, 368). Countering sustainability wrongdoings as a key responsibility of all stakeholders engaged into this process together with other internal corporate compliance processes should be properly monitored. There are many examples of sustainability wrongdoings as for example greenwashing practices that thanks to measurable indicators can be monitored. Also other tools of wider compliance concept can be used in the process of sustainability enforcement (Baldwin & Cave, 2021, 82). Additionally various compliance incentive instruments can be useful (Gargantini, 2019, 291). Materiality is the criterion through which companies categorise events included in sustainability reports to satisfy both regulatory requirement and expectations of stakeholders. Application of materiality criterion in sustainability reports is one of the tools to support stakeholders' engagement. It differs among industries that engage different groups of actors and applies Global Reporting Initiative and International Integrated Reporting Council guidelines that both assume their involvement. The statistical analysis highlights the importance of industry, Global Reporting Initiative Standards application and stakeholder engagement in the reporting process, in the materiality analysis, that helps to achieve an augmented level of materiality application and enhanced report quality (Torelli et al., 2019).

The absence of extensive mandatory regulation and auditing, differences in corporate governance structures and dynamics result in variations in choices concerning sustainability reporting and the underlying corporate sustainability performance. Monitoring of board effectiveness has an impact on governance mechanism and is positively corelated with sustainability reporting quality, compliance with stainability reporting laws and regulations, and sustainability performance itself (Manning et al., 2018).

Enforcement of sustainability policies is a multilevel and multi-partite process. It is multilevel as the laws and regulations are being introduced on various levels from international to very local. It is multi-partite as establishment, implementation and enforcement of these laws and regulations engage various groups of stakeholders (Plater, 2006, 511–521). In this sense multicentric and multilevel governance are tested and useful tools that involve multiple stakeholders. The advantage of multilevel and multicentric characteristics of the enforcement process is that it takes into account the specificity of the jurisdictions in which they take place. Efficiency of the enforcement may be affected dependent on country legal tradition and other variables as complexity of the matter, sector in which it operates, activeness of regulators etc. (Fernandez-Feijoo et al., 2015).

Companies themselves can be engaged into the process of enforcement of the sustainability laws, regulations and policies. Enforcement of sustainability with an active role of companies is based on the mechanism of controls of a way that they implement their internal sustainability policies that due to their global reach and impact can have a wider effect. This indirectly refers to the concept of sustainable corporate governance (World Economic Forum, 2021, 26). Although sustainability as a catchall term refers to multiple areas however, companies address sustainability issues more effectively when they combine missions and long-term plans in the way that they focus on particular sustainability topics. The specific sustainability topics being in line with the worldwide companies' priorities have a greater chance for efficient enforcement. Once the companies define the sustainability areas that matter for them, they become impactful enforcement agents with broad coverage. Their efficiency is the outcome of the fact these policies are beneficial to the business gives a competitive advantage. Whichever commercial motivation backs the decision to enforce sustainability, the involvement of the growth driven corporations is useful (McKinsey, 2021).

Global businesses' compliance programmes are based on engagement of all stakeholders that interact with them. Multi-jurisdictional corporates together with public authorities create an enforcement system for sustainable laws, regulations and policies. Additionally, sustainability conduct requirements introduced at various levels constitute criteria for assessment of an extent to which managements of the businesses are sustainability compliant. Institutional investors can have their role in this process as well. Assessment to what extend they are interested in financing proposed products and technologies depends on their sustainability and constitute an element of the enforcement. They use various criteria to assess it and obligate corporates to report it. The control over compliance with sustainability conduct is still a relative novelty of sustainability assurance. Selection of right criteria for sustainability assessment is an ongoing subject of discussions. There are however six criteria that are discussed the most: human and environmental well-being, competence about facts and values, fairness in process and outcome, a reliance on human strengths rather than weaknesses, the opportunity to learn and efficiency. It is hoped that an explicit discussion of the appropriate criteria for environmental decisions will lead to better investment decisions and create efficient sustainability enforcement tools (Dietz, 2003, 33–39).

Unification of Sustainability Enforcement Tools

Regardless of the role of corporates in the sustainability enforcement process this is on the regulators that should lead the path of designing innovative approach to achieve efficiency of the laws, regulations, and policies and consider a methodic approach to consolidating them. As much as diversified the sustainability problems are, the same applies to any attempt of regulating them. The richness of measures reflected in multicentric and multilevel legislation at international, national, and subnational levels refers to variety of sectors and industries impacted by environmental challenges. This creates a complex network of interrelated regulations that either already exist or need to be created.

Many of the measures in the hands of lawmakers and regulators are *per se* a challenge to limitations of a traditional approach to normative landscape. They require an openminded view on the role of the sustainability regulation and an engagement of multiple actors on various levels. Some examples of these measures have been identified within the EU initiatives. The area of sustainability and in particular such problems as climate change, renewable energy, societal inequalities, environmental justice, fishery quotas, ecological agriculture, wildlife protection etc. are in the centre of the EU agenda (Wenta et al., 2019, 109). To address them the EU established specialised agencies equipped with the enforcement tools such as the Committee of the Regions. This subsidiary body of the EU has been established to support multilevel policies' enforcement at the local and regional levels. In line with these promotional undertakings, the Committee of the Regions adopted a Charter for Multilevel Governance in Europe calling public authorities of all levels to use and promote the multileveled governance (Och, 2020, 05). The EU Cohesion Policy with almost a third of the total EU budget and covering its sustainability initiatives increases ownership on vertical and horizontal levels of governance. It emphasises the importance of multilevel governance (Dobravec et al., 2021, 11).

Sustainability enforcement referencing driven by investors' prioritise impact lending developed to a broad range such as: achievement of zero net carbon emissions or percentage reductions in carbon emissions; commissioning and achieving a sustainability accreditations by a rating agencies; introducing policies that refer to monitoring of environmental, social and governance indicators; assessment of impact of supply chains for human and employees rights abuses, slavery and poor labour standards; maintaining appropriate corporate codes of ethics; participating in initiatives to improve local communities etc (Siri & Zhu, 2019, p. 4).

Enforcement tools that become growingly important are the ones related to new technologies. They reflect and shape the societal changes and can enhance sustainability enforcement more effectively than any other tools. Another advantage is that new technologies are able to address sustainability challenges in the way that they can help to identify and mitigate potential adverse impacts. The technological changes enhance societal, organisational, political, and economic shift that takes place in various sectors, such as energy generation, countering inequalities, logistics and supply chains, water management, circular economy etc. For example, a carbon-free technologies may require establishment of new value chains. It may mean a long process that can alter society in several ways by changing consumer behaviours, distributional effects, hard infrastructural solutions and establishing new business networks (Soderholm, 2020, 5).

The societal acceptance expressed by the grassroots support as a result of a growing European Environmental Conscience for the novel policies implementation tools and its openness for direct dialogue with regulators derives from their awareness of the existential threat for the planet related to anthropogenic roots of the damages for the environment and the complexity the of problem how to systemically respond to this threat (Kuenzer, 2024, 186). Environmental policies vary even within the EU despite of the fact it is the organisation strongly unified in many other spheres (Roelfsema et al., 2020, 2096). However, enforcement of environmental policies requires close and orchestrated cooperation globally. Inefficiency of inward-looking policies misses the opportunity of public engagement. Given the fact that properly implemented multilevel enforcement of environmental policies, that are an important part of the sustainability agenda, could play an integrating and economy boosting role, especially the use of financial tools deserves to be thoroughly explored.

Global sustainability challenges by definition are cross-border, therefore their enforcement tools need to be global as well. To be efficient they also need to be smart and adjusted to the changing societal, economic, cultural, and political reality (Weiss & Kammel, 2015, XX). Contemporary global complexity caused by human or nature have made this need even more apparent. The task ahead of regulators and lawmakers is to engage with all stakeholders. This engagement means learning from tested patterns and sharing ownership but may lead to reducing conflicts, encouraging innovation in finding solutions. It can foster inclusive decision making, promoting equity, enhancing decentralised decision making, and building social capital (Kartadjumena & Rodgers, 2019, 1673). Stakeholders engagement has proven to be efficient in other spheres it may be so in case of sustainability, too (Manetti, 2011, 112). Engagement can also be an opportunity from the perspective of societal processes when stakeholders mutually learn their values, reflect and establish a common set of them. Dialogue is useful in increasing awareness, changing attitudes, and affecting behaviours. There is a need for an approach that enhances ethical perspectives of sustainability. An open dialogue approach to sustainability policies provides means to enforce the policies and laws (Narin Mathur et al., 2008, 601–609).

Conclusions

The efficient sustainability norms and policies issuance is undeniably a complex and multifaceted process. The enforcement of them as a next phase of this process for its efficiency needs to refer to multilevel tools including the financial ones. The discussion on how to assure sustainability enforcement referred to in this chapter regards financial and normative instruments, with a focus on multilevel governance. This text synthesises these discussions, emphasising the vital integration of sustainability as a core principle within global regulatory frameworks, inspired by the financial industry's experiences. The multilevel approach to sustainability enforcement is not just a legislative strategy but also a necessity. The interdependence of global systems has an impact on a governance structure that transcends national boundaries and sectoral divides. This approach allows for the alignment of policies across different governance levels – international, national, and local – ensuring that sustainability goals are pursued in a coherent and coordinated fashion. Such alignment is critical to face the challenges of sustainability, such as climate change, biodiversity loss, and resource is transboundary.

The use of multilevel enforcement tools, i. e. these that engage various private and public stakeholders, is a regulatory trend that has already been tested in various sectors. By engaging businesses and other organised society actors it assures a bottom-up approach to sustainability policies implementation (Smith, 2008, 361). On the other side the top-down financial measures used by multi-layered regulatory powers like lawmakers and public administration means the system needs to be hybrid and continue deploying a more traditional one (Eicken et al., 2021, 473). The financial sector's experience within stringent regulatory frameworks provides useful lessons for sustainability enforcement. The aftermath of the global financial crises exemplified how multilevel regulatory measures could strengthen system resilience. Similarly, sustainability challenges require robust, transparent, and enforceable policies that are coordinated across various governance levels to mitigate risks and capitalise on interdependencies.

Financial instruments have emerged as pivotal tools in the sustainability enforcement. Instruments like green bonds, carbon pricing mechanisms, and sustainability-linked loans demonstrate how financial tools can align market operations with environmental outcomes. The success of these instruments in the financial sector underscores the potential for similar mechanisms to drive sustainability policies. Green bonds, for instance, have provided a blueprint for raising capital for environmental projects at scale. These bonds highlight how regulatory frameworks can incentivise investments in sustainability while offering financial returns. This dual-benefit mechanism is crucial for engaging private sector participants who might otherwise be reluctant to divert resources towards sustainable projects without a clear economic incentive.

Despite of a *sui generis* subsidiarity left to countries that may select their own ways to assure sustainability enforcement there is a need for unification of sustainability enforcement measures as the sustainability is a global problem and the risks associated with it are global. Particular benefits for efficiency of the sustainability enforcement programmes derives from their interrelations with corporate practices. Corporate normative instruments, including soft law mechanisms like guidelines, codes of conduct, and voluntary standards, play a critical role in sustainability enforcement. These tools are particularly effective in areas where hard law might be too rigid or slow to implement. The flexibility of soft law allows it to adapt rapidly to emerging sustainability challenges and to foster innovation in corporate governance. Corporate governance itself, influenced heavily by normative instruments, has seen a shift towards sustainability integration. The principles of ESG criteria have begun to reshape investment decisions and corporate strategies globally. This shift is reflective of a broader trend where sustainability is increasingly viewed as integral to long-term profitability and risk management, rather than a peripheral concern.

The experience of financial governance in implementing complex and effective regulatory frameworks offers critical lessons for sustainability policies enforcement. Financial regulators have developed a sophisticated understanding of how to balance stakeholder interests, enforce regulations, and ensure compliance. Some regulatory techniques used to monitor and enforce compliance in the financial sector, such as stress testing and risk assessment frameworks, can be adapted to monitor sustainability impacts. Such adaptations could enhance the effectiveness of sustainability policies, ensuring that they are not only well-crafted but also robustly enforced. These insights are therefore directly applicable to the field of sustainability, where enforcement challenges are similarly complex and multifaceted. It refers especially to sustainability policies implementation lessons taken from the financial industry which is crucial from the viewpoint of investment allowing for introducing new technologies and improving the existing ones. Although multilevel sustainability enforcement tools may evoke numerous ethical dilemmas, the corporate lessons and specifically the ones taken from financial industry may support enforcement of sustainability polices in efficient manner (Crane et al., 2019, 233). The principles of financial regulation provide a strong foundation upon which sustainability policies can be built in the future. This have to be characterised by innovative regulatory approaches, stakeholder engagement, and continuous learning from prior successes and failures of similar regulatory processes.

Notes

1 Regulation (EU) 2018/842 of the European Parliament and of the Council of 30 May 2018 on binding annual greenhouse gas emission reductions by Member States from 2021 to 2030 contributing to climate action to meet commitments under the Paris Agreement and amending Regulation (EU) No 525/2013 (Text with EEA relevance, *OJ L 156, 19.6.2018*).

2 Regulation (EU) 2018/841 of the European Parliament and of the Council of 30 May 2018 on the inclusion of greenhouse gas emissions and removals from land use, land use change and forestry in the 2030 climate and energy framework, and amending Regulation (EU) No 525/2013 and Decision No 529/2013/EU (Text with EEA relevance).

3 Regulation (EU) 2019/2088 of the European Parliament and of the Council of 27 November 2019 on sustainability-related disclosures in the financial services sector (Text with EEA relevance, *OJ L 317, 9.12.2019*).

4 Regulation (EU) 2019/2089 of the European Parliament and of the Council of 27 November 2019 amending Regulation (EU) 2016/1011 as regards EU Climate Transition Benchmarks, EU Paris-aligned Benchmarks and sustainability-related disclosures for benchmarks (Text with EEA relevance, *OJ L 317, 9.12.2019*).

5 Regulation (EU) 2020/852 of the European Parliament and of the Council of 18 June 2020 on the establishment of a framework to facilitate sustainable investment, and amending Regulation (EU) 2019/2088 (Text with EEA relevance, *OJ L 198, 22.6.2020*).

6 Regulation (EU) 2023/2631 of the European Parliament and of the Council of 22 November 2023 on European Green Bonds and optional disclosures for bonds marketed as environmentally sustainable and for sustainability-linked bonds (*OJ L, 2023/2631, 30.11.2023*).

7 Directive (EU) 2022/2464 of the European Parliament and of the Council of 14 December 2022 amending Regulation (EU) No 537/2014, Directive 2004/109/EC, Directive 2006/43/EC and Directive 2013/34/EU, as regards corporate sustainability reporting (Text with EEA relevance, *OJ L 322, 16.12.2022*).

Bibliography

Akbar, M. W., Yuelan, P., Zia, Z., & Arshad, M. I. (2021). Role of fiscal policy in energy efficiency and CO_2 emission nexus: An investigation of belt and road region. *Journal of Public Affairs, 21*(2), 2605.

Alexander, K. (2022). Global financial governance and banking regulation: Redesigning regulation to promote stakeholder interests. In J. Pauwelyn, M. Maggeti, T. Buthe, & A. Berman (Eds.), *Rethinking participation in global governance: Challenges and reforms in financial and health institutions* (pp. 4–25). Oxford University Press.

Baldwin, R., & Cave, M. (2021). *Taming the corporation: How to regulate for success.* Oxford University Press.

Barnes, P. M., & Hoerber, T. C. (Eds.). (2013). *Sustainable development and governance in Europe: The evolution of the discourse on sustainability.* Routledge.

Bastedas-Orteaga, E., & Stewart, M. G. (2019). *Climate adaptation engineering: Risks and economics for infrastructure decision-making.* Butterworth-Heinemann.

Bellucci, M., & Manetti, G. (2018). *Stakeholder engagement and sustainability reporting.* Routledge.

Benz, A. (2021). *Policy change and innovation in multilevel governance.* Edward Elgar Publishing.

Braun, T. (2019). The quasi-legislative measures of international corporations. *International Journal of Legal Studies, 2*(6), 123–145.

Busch, D., Ferrarini, G., & Grünewald, S. (2021). *Sustainable finance in Europe.* Springer.

Choy, S., Jiang, S., Liao, S., & Wang, E. (2022). Public environmental enforcement and private lender monitoring: Evidence from environmental covenants. *Rotman School of Management Working Paper, 3860178, 10*(1), 1–30.

Colgan, J., Green, J., & Hale, T. (2021). Asset revaluation and the existential politics of climate change. *International Organization, 75*(2), 588–620.

Crane, A., Matten, D., Glozer, S., & Spence, L. J. (2019). *Business ethics: Managing corporate citizenship and sustainability in the age of globalization.* Oxford University Press.

de Oliveira Neves, R. (2022). The EU taxonomy regulation and its implications for companies. In P. Camara & F. Morais (Eds.), *The Palgrave handbook of ESG and corporate governance* (pp. 1–25). Palgrave Macmillan.

De Smet, Aaron, Gao, Wenting, Henderson, Kimberly, Hundertmark, Thomas (2021), Organising for sustainability success: Where, and how, leaders can start, McKinsey&Company Report, https://netzerohub.id/wp-content/uploads/2022/02/organizing-for-sustainability-success-where-and-how-leaders-can-start.pdf

Dietz, T. (2003). What is a good decision? Criteria for environmental decision making. *Human Ecology Review, 10*(1), 33–39.

Dobravec, V., Matak, N., Sakulin, C., & Krajacic, G. (2021). Multilevel governance energy planning and policy: A view on local energy initiatives. *Energy, Sustainability and Society, 11*(2), 11.

Eicken, H., Danielsen, F., Sam, J.-M., Fidel, M., Johnson, N., Poulsen, M. K., Lee, O. A., Spellman, K. V., Iversen, L., Pulsifer, P., & Enghoff, M. (2021). Connecting top-down and bottom-up approaches in environmental observing. *BioScience, 71*(5), 473–487.

Eyl-Mazzega, M. A., & Mathieu, C. (2019). Strategic dimensions of the energy transition: Challenges and responses for France, Germany and the EU. *Etudes de l'IFRI,* 25–47.

Fernandez, R. (2021). Community renewable energy projects: The future of the sustainable energy transition? *The International Spectator, 56*(1), 94–112.

Fernandez-Feijoo, B., Romero, S., & Ruiz, S. (2015). Multilevel approach to sustainability report assurance decisions. *Australian Accounting Review, 25*(4), 346–358.

Fiorino, D. J. (2018). *Can democracy handle climate change?* Polity.

Ganguly, G., Setzer, J., & Heyvaert, V. (2018). If at first you don't succeed: Suing corporations for climate change. *Oxford Journal of Legal Studies, 38*(4), 852–878.

Gargantini, M. (2019). Corporate governance, financial information and EU market abuse regulation. In D. Busch, G. Ferrarini, & G. van Solinge (Eds.), *Corporate governance of financial institutions* (pp. 291–315). Oxford University Press.

Gargantini, M., & Di Noia, C. (2020). The approval of prospectus: Competent authorities, notification and sanctions. In D. Busch, G. Ferrarini, & J. P. Franx (Eds.), *Prospectus regulation and prospectus liability* (pp. 370–398). Oxford University Press.

Ghoogossian, C. (2015). Evading the transparency tragedy: The legal enforcement of corporate sustainability reporting. *Hastings Business Law Journal, 11*(2), 368–397.

Glavas, C. & F. J. Okala Onana (2025). The EU Green Bond Market. In M. Tendero & C. Weber (Eds.), *The European Environmental Conscience in the EU (forthcoming)*. Routledge.

Grolleau, G., Mzoughi, N., Peterson, D., & Tendero, M. (2022). Changing the world with words? Euphemisms in climate change issues. *Ecological Economics, 193*, 107307.

Grolleau, G., & Weber, C. S. (2024). The effect of inflation on CO_2 emissions. *Ecological Economics, 217*, 108029.

Heddeman-Robinson, M. (2018). *Enforcement of international environmental law: Challenges and responses at the international level*. Routledge.

Hill, J. (2020). *Environmental, social, and governance investing: A balanced analysis of the theory and practice of a sustainable portfolio*. Academic Press.

Hoerber, T. (2012). *The origins of energy and environmental policy in Europe: The beginnings of a European environmental conscience*. Routledge.

Hoerber, T., Weber, G., & Giraud, A. (2021). Towards ego-ecology? Populist environmental agendas and the sustainability transition in Europe. *The International Spectator, 56*(1), 51–67.

Kartadjumena, E., & Rodgers, W. (2019). Executive compensation, sustainability, climate, environmental concerns, and company financial performance: Evidence from Indonesian commercial banks. *Sustainability, 11*(6), 1673.

Knodt, M. (2019). Multilevel coordination in EU energy policy: A new type of "harder" soft governance? In N. Behnke, J. Broschek, & J. Sonnicksen (Eds.), *Configurations, dynamics, and mechanisms of multilevel governance* (pp. 181–203). Springer.

Kuenzer, J. (2024). A growing European environmental conscience. In T. Hoerber & G. Weber (Eds.), *The European environmental conscience in EU politics: A developing ideology* (pp. 186–205). Routledge.

Lehman, M. (2020). EU law-making 2.0: The prospect of a European business code. *European Review of Private Law, 28*(1), 75–99.

Lepore, L., & Pisano, S. (2022). *Environmental disclosure: Critical issues and new trends*. Routledge.

Macchiavello, E., & Siri, M. (2020). Sustainable finance and fintech: Can technology contribute to achieving environmental goals. A preliminary assessment of 'green fintech'. *European Banking Institute Working Papers, 71*, 130.

Machiavello, E., & Siri, M. (2022). Sustainable finance and fintech: Can technology contribute to achieving environmental goals? A preliminary assessment of 'green fintech' and 'sustainable digital finance'. *European Company and Financial Law Review, 19*(1), 6–35.

Maltais, A., & Nykwist, B. (2020). Understanding the role of green bonds in advancing sustainability. *Journal of Sustainable Finance and Investment, 10*(1), 1–23.

Manetti, G. (2011). The quality of stakeholder engagement in sustainability reporting: Empirical evidence and critical points. *Corporate Social Responsibility and Environmental Management, 18*(2), 112–131.

Manning, B., Bramm, G., & Reimsbach, D. (2018). Corporate governance and sustainable business conduct—Effects of board monitoring effectiveness and stakeholder engagement on corporate sustainability performance and disclosure choices. *Corporate social responsibility and environmental management*. Wiley Online.

Mathur, V. N., Price, A. D. F., & Austin, S. (2008). Conceptualizing stakeholder engagement in the context of sustainability and its assessment. *Construction Management and Economics*, *16*(6), 601–609.

Millar, H., Bourgeois, E., Bernstein, S., & Hoffmann, M. (2021). Self-reinforcing and self-undermining feedbacks in subnational climate policy implementation. *Environmental Politics*, *30*(5), 887–905.

O'Farrell, P. J., & Anderson, P. M. L. (2010). Sustainable multifunctional landscapes: A review to implementation. *Current Opinion in Environmental Sustainability*, *2*(1–2), 62–66.

Och, M. (2020). Sustainable finance and the EU taxonomy regulation – Hype or hope? *KU Leuven Working Papers*, *5*, 1–15.

Olah, J., Aburumman, N., Popp, J., Asif Khan, M., Haddad, H., & Kitukytha, N. (2020). Impact of Industry 4.0 on environmental sustainability. *Sustainability*, *12*(11), 4567.

Omarova, S. T. (2020). Technology vs. technocracy: Fintech as a regulatory challenge. *Journal of Financial Regulation*, *6*(1), 78–104.

Pereira, L., Asrar, G. R., Bhargava, R., Fisher, L. H., Hsu, A., Jabbour, J., Nel, J., Selomane, O., Sitas, N., Trisos, C., Ward, J., van den Ende, M., Vervoort, J., & Weinfurter, A. (2021). Grounding global environmental assessments through bottom-up futures based on local practices and perspectives. *Sustainability Science*, *16*(6), 1918–1935.

Plater, Z. J. B. (2006). Law, media and environmental policy: A fundamental linkage in sustainable democratic governance. *B.C. Environmental Law Review*, *33*(2), 511–521.

Quaglia, L. (2020). *The politics of regime complexity in international derivatives regulations*. Oxford University Press.

Rishi, P. (2022). Behavioural transformation for sustainability and pro-climate action. In *Managing climate change and sustainability through behavioural transformation: Sustainable development goals series* (pp. 150–171). Springer Nature.

Roelfsema, M., van Soest, H. L., Harmsen, M., et al. (2020). Taking stock of national climate policies to evaluate implementation of the Paris Agreement. *Nature Communications*, *11*, 2096.

Rogers, D. T. (2022). *Environmental compliance handbook: Sustainability and future environmental regulations*. Routledge.

Rossinioli, F., Stacchezzini, R., & Lai, A. (2021). Financial analysts' reaction to voluntary integrated reporting: Cross-sectional variation in institutional enforcement contexts. *Journal of Applied Accounting Research*, *23*(1), 32–52.

Sachs, J. D., Woo, W. T., Yoshino, N., & Taghizadeh-Hesary, F. (Eds.). (2020). Importance of green finance for achieving sustainable development. In *Handbook of green finance. Energy, security and sustainable development* (pp. 7–32). Springer Nature.

Sarker, T., Taghizadeh-Hesary, F., Mortha, A., & Saha, A. (2020). The role of fiscal incentives in promoting energy efficiency in the industrial sector: Case studies from Asia. *Asian Development Bank Institute Working Paper*, *1172*(08), 1–20.

Sheehy, B. (2019). TNC code of conduct or CSR? A regulatory systems perspective. In M. M. Rahim (Ed.), *Code of conduct on transnational corporations: Challenges and opportunities* (pp. 80–95). Springer.

Siri, M., & Zhu, S. (2019). Will the EU commission successfully integrate sustainability risks and factors in the investor protection regime? A research agenda. *Sustainability*, *11*(22), 6292.

Smith, J. L. (2008). A critical appreciation of the "bottom-up" approach to sustainable water management: Embracing complexity rather than desirability. *Local Environment*, *13*(4), 361–375.

Soderholm, P. (2020). The green economy transition: The challenges of technological change for sustainability. *Sustainable Earth, 3*(6), 5–22.

Sugiawan, Y., & Manag, S. (2019). Public acceptance of nuclear power plants in Indonesia: Portraying the role of a multilevel governance system. *Energy Strategy Reviews, 26,* 9–20.

Taghizadeh-Hesary, F., & Yoshino, N. (2020). Sustainable solutions for green financing and investment in renewable energy projects. *Energies, 14*(4), 788.

Tanaka, K. (2011). Review of policies and measures for energy efficiency in industry sector. *Energy Policy, 39*(10), 6547–6556.

Torelli, R., Balluchi, F., & Furlotti, K. (2019). The materiality assessment and stakeholder engagement: A content analysis of sustainability reports. *Corporate social responsibility and environmental management.* Wiley Online.

Vantagiatto, F. P. (2020). Regulatory relationships across levels of multilevel governance systems: From collaboration to competition. *Governance, 33*(2), 183–204.

Waas, T., Hugé, J., Block, T., Wright, T., Benitez-Capistros, F., & Verbruggen, A. (2014). Sustainability assessment and indicators: Tools in a decision-making strategy for sustainable development. *Sustainability, 6*(9), 5530–5555.

Weber, C. (2025). Going Green? The Changing Role of the ECB in the sustainability Strategy of the EU. In M. Tendero & C. Weber (Eds.), The European Environmental Conscience in the EU (forthcoming). Routledge.

Weiss, F., & Kammel, A. J. (Eds.). (2015). *The changing landscape of global financial governance and the role of soft law.* Brill.

Wenta, J., McDonald, J., & McGee, J. S. (2019). Enhancing resilience and justice in climate adaptation laws. *Transnational Environmental Law, 8*(1), 109–137.

World Economic Forum. (2021, June 26). The EU wants business to be sustainable. But it must empower companies to do that.

Zeitlin, J. (2023). Hierarchy, polyarchy, and experimentalism in EU banking regulation: Single supervisory mechanism in action. *Journal of European Integration, 45*(1), 79–101.

Ziolo, M., Filipiak, B. Z., & Tundys, B. (2021). *Sustainability in bank and corporate business models.* Springer.

5   The EU Green Bond Market

Dejan Glavas and Florent Junior Okala Onana

Introduction

The international commitment to address climate change, particularly reflected in the Paris Agreement, has prompted significant shifts in global finance. In particular, the green bond market has experienced a swift ascendancy, becoming a vital instrument for financing projects with positive environmental and climate benefits. The green bond market's growth underscores the emergent alliance between environmental sustainability and financial practices. This phenomenon is more specifically visible within the European Union (EU), which has set robust climate targets such as achieving carbon neutrality by 2050 and is poised to lead in sustainable finance. An essential part of this strategy involves harnessing the power of capital markets, which requires an annual additional investment of around €480 billion over the next decade. In parallel, regulations such as the EU's Taxonomy Regulation and the Corporate Sustainability Reporting Directive (CSRD) have also been instrumental in providing frameworks for sustainable financial practices.

This chapter aims to provide an in-depth examination of the green bond market and related sustainability financial mechanisms within the EU, applying the European environmental conscience methodology (Hoerber & Weber 2021). This approach allows us to examine how green bonds connect energy and environmental policies, represent an innovative sustainable finance solution, demonstrate European Commission expertise and leadership, and reflect grassroots demand for green investments. This analysis will begin with a clear definition and discussion of green bond standards, which are essential for understanding the market's growth and regulatory needs.

Guided by this context, this research will explore the following questions:

- How has the Paris Agreement influenced the development and growth of the green bond market?
- What role do regulations like the EU's Taxonomy Regulation, the EU Green Bond Standard (EuGB) and the proposed CSRD play in the trajectory of the green bond market?

DOI: 10.4324/9781032656359-8

- What challenges and ambiguities persist within the green bond market, and how might they be addressed to optimise the market's contribution to sustainability objectives?

Based on our preliminary literature review, we have formulated several hypotheses. Firstly, we posit that the Paris Agreement has functioned as a significant catalyst for the growth of the green bond market, particularly in regions characterised by robust regulatory frameworks such as the EU. This is supported by evidence showing increased green bond issuance post-Agreement, linked to enhanced investor confidence and regulatory support (e.g., EU Taxonomy Regulation). Secondly, we anticipate that regulatory measures have played an instrumental role in shaping the green bond market. For instance, the EU Taxonomy Regulation and the CSRD have been key in fostering transparency and standardisation, significantly enhancing investor confidence. Market data shows a correlation between these regulations and increased green bond issuance. Finally, our third hypothesis acknowledges the considerable strides made in the green bond market. However, it also highlights the persisting challenges, notably those related to market ambiguities and the effectiveness of standards enforcement. We propose that these unresolved issues could significantly influence the sustainability outcomes of these financial instruments.

In conclusion, the intersection of international climate commitments, sustainable finance, and regulatory frameworks holds the key to achieving global sustainability goals. This chapter aims to comprehensively explore this junction, with a special emphasis on the green bond market within the EU. We anticipate that the insights gleaned from this study will not only expand our understanding of the complex dynamics governing the green bond market, but also inform policy and investment decisions that could accelerate the transition towards a sustainable economy. With the hypotheses derived from our preliminary literature review guiding our investigation, we will critically analyse the drivers of the green bond market, the potential impact of regulatory measures such as the EuGB, and the unresolved challenges that could affect market efficiency and the achievement of sustainability objectives. Through this endeavour, we hope to contribute valuable insights to the discourse on sustainable finance.

Green Bond Definition and Standards Debate

Green Bond Definition

We build on the definition of green bonds from the International Capital Markets Association (ICMA) to explain how the green bond market took off and how it specifically developed within the EU market. It is important to underline that this definition and the green bond principles are voluntary guidelines and are by no means legally binding. According to the definition :
Green Bonds are any type of bond instrument where the proceeds or an equivalent

amount will be exclusively applied to finance or re finance in part or in full, new and/or existing eligible Green Projects and which are aligned with the four core components of the Green Bond Principles (ICMA, 2018). The four components include use of proceeds, process for project evaluation and selection, management of proceeds and reporting. This definition has structured the green bond market for several reasons.

First, it has focused on the main distinguishing characteristic of the green bond market, which is the use of proceeds, while such use of proceeds is usually for general corporate purposes. This distinction is fundamental since it guarantees that there is no fungibility of funds for green bonds contrary to standard bonds.

The second characteristic concerns the fact that the funds will be used to finance or refinance projects. The definition leaves the door open to the possibility of refinancing existing projects rather than dedicating green bond issuances only to new projects. This characteristic raises the question of additionality. The underlying question here is to know whether only new projects are helping our economies to reduce their carbon footprint or whether refinancing existing projects helps as well. The main argument in the academic community for defending this position is that refinancing green projects can help firms to obtain better financing conditions thanks to the use of green bonds. These conditions would then incentivise firms to use the funds to further invest in new green projects. Research found that green bonds offer better financing conditions for firms, more specifically from the primary market (Kapraun et al., 2021), supporting this argument. The academic debate still exists whether to know how green bonds really affect the environmental impact of firms (Flammer, 2021). The question of a use of green bonds that is not primarily driven by funding green projects may also be asked (Glavas, 2022).

The third point of focus of the definition concerns the fact that projects may be financed in part or in full. This third section of the definition leads to the question of attribution of benefits from green bonds. Which part of the financed project's environmental benefits can be attributed to the green bond as compared to the rest of the financing? Attributing the benefits to the share of the financing would not be fully rigorous as it may be the case that the project would not have existed if the green funding had not been provided. Otherwise, if the full project could have been funded with traditional finance, the question of additionality could then be asked again.

Finally, the last fundamental element of the definition is that green bonds can be used to fund eligible green projects. This section of the definition leads to an understanding that the green bond market is not closed to firms evolving in non-green sectors if the project being financed is green. This part of the definition also anticipates that there would be the need for firms that are not in green sectors, that these would need to improve their processes to produce their products or provide their services for them to be aligned with a greener future.

To summarise, the definition of green bonds has been material in shaping the market from its premises. It has created a large framework allowing a wide array of issuers to enter the market. Indeed, the definition does not close the door to firms that wish to refinance existing green projects, which mix green and non-green bonds to fund projects and that are coming from unsustainable sectors. The definition, by its very non-binding nature pleaded for a green bond market based on soft rather than hard law, meaning that it would rely on rules that are not compulsory. By keeping the door open, this definition has contributed to the expansion of the green bond market.

While the ICMA definition has been instrumental in shaping the green bond market, recent developments in the EU have sparked a debate about the need for more standardised and enforceable green bond criteria.

Green Bond Standards Debate

The debate surrounding green bond standards, particularly the EuGB, underscores the complexity of sustainable finance regulation. As the green bond market grows rapidly, the need for robust and clear standards becomes increasingly key to ensure that investments truly contribute to environmental sustainability. The proposed EuGB aims to enhance transparency and comparability in the green bond market, addressing some of the existing market deficiencies. Pyka (2023) argues that while the EuGB represents a significant step forward, it may not go far enough in ensuring private enforcement. This concern is echoed by other researchers who stress the importance of robust enforcement mechanisms to prevent greenwashing, where bonds are falsely marketed as green (Freeburn & Ramsay, 2020).

The success of the EuGB hinges on the development of rigorous enforcement mechanisms. Greenwashing poses a serious threat to the credibility of the green bond market, as it can mislead investors and undermine the environmental benefits that these financial instruments are supposed to deliver. Freeburn and Ramsay (2020) emphasise that without stringent oversight and enforcement, the integrity of the green bond market could be compromised. They argue that enforcement should not only be robust but also adaptable to the evolving nature of green finance.

Marcos and Castrillo (2022) highlight the potential of the EuGB to become a global benchmark for green bond standards. However, they also note significant challenges in its implementation and verification. Ensuring that green bonds adhere to the proposed standards requires broad monitoring and reporting frameworks. This includes third-party verification and the establishment of clear criteria for what constitutes a green bond. The authors suggest that international cooperation and the harmonisation of standards are critical to overcoming these challenges.

The necessity for international compatibility and stringent standards to bridge the gap between national and international frameworks is evident in China's approach to green bond standards (Liang & Gao, 2022). China's Catalogue of Green Bond Support Projects (People's Bank of China, 2021) provides a unified classification for green bonds, aiming to avoid discrepancies and facilitate international investment. This approach underscores the global effort to standardise green finance and highlights the importance of aligning national standards with international expectations.

The high demand for green bonds creates pressure to relax standards, which could undermine their credibility. Freeburn and Ramsay (2020) discuss the potential dangers of this trend, noting that relaxing standards to meet demand can lead to green defaults, where the promised environmental benefits are not delivered. Such outcomes can damage investor confidence and reduce the overall impact of green bonds. To mitigate this risk, it is important to maintain rigorous standards even in the face of high market demand.

The rapid expansion of the green bond market necessitates rigorous standards to maintain investor confidence and ensure genuine environmental impact. Hoinaru et al. (2020) discuss the EuGB's potential to foster a more cohesive and efficient green bond market within the EU and beyond. They emphasise the need for clear, enforceable regulations that can adapt to market growth while maintaining the integrity of green bonds. A robust regulatory framework is essential for sustaining investor trust and achieving long-term environmental goals.

The development of a unified classification for green bonds, as seen in China's Catalogue of Green Bond Support Projects (2021), highlights the global efforts to standardise green finance. Such classifications help to avoid discrepancies and facilitate international investment, making it easier for investors to identify genuinely green projects. This effort towards standardisation reflects a broader trend in the financial industry towards greater transparency and accountability in sustainable finance.

Overall, the debate surrounding the EuGB and green bond standards underscores the ongoing evolution of the EU's approach to sustainable finance regulation. Balancing market growth, integrity, and environmental impact remains a complex challenge. While the EuGB represents a significant step forward, continuous efforts are needed to enhance enforcement mechanisms, harmonise international standards, and maintain the credibility of green bonds. As the green bond market continues to expand, the development of robust, transparent, and enforceable standards will be essential to ensuring that these financial instruments deliver genuine environmental benefits and contribute to global sustainability goals.

Having established the definition and standards for green bonds, it is key to understand the motivations behind their issuance. The following section explores the numerous factors that drive firms to enter the green bond market.

Why Do Firms Issue Green Bonds?

In 2022, according to Climate Bond Initiative data, green bond issuance experienced a year-over-year drop for the first time in a decade, amounting to USD 487.1 billion, a decline of 16% compared to 2021 (Climate Bond Initiative, 2022). Despite this decrease, green bonds maintained 3% of overall issuance volumes.

The green label still dominated the global thematic debt issuance, accounting for 56% of the total. Demand for green bonds continued to outpace supply, supported by investors with green or socially responsible inclinations. Several nations allocated a portion of their fiscal spending to expedite the shift to a low-carbon economy, with France being a prominent issuer. The global energy crisis intensified clean energy policies, with capital being allocated towards boosting renewable energy capacity and developing green technologies.

In 2022, 382 new issuers accounting for USD 142 billion of green volumes joined the market (CBI, 2022). Non-financial corporates were responsible for 37% of the debut issuer volume. Developed markets contributed 67% of the 2022 green bond volume, while emerging markets and supranational issuers contributed 23% and 9%, respectively. The European Commission's green bond programme significantly contributed to the market growth, aiming to fund 30% of its NextGenerationEU recovery plan through dedicated green bonds. The Asia-Pacific region, led by China, Japan, and South Korea, also saw substantial green bond issuance, primarily due to the increasing role of financial corporate issuers.

The burgeoning popularity of green bonds in the financial markets is a clear signal of a shift towards environmental consciousness. The Paris Agreement, a landmark in the fight against climate change, led to an observable surge in the green bond market. Yet, this surge cannot be attributed to financial gains alone; a key driver of this trend is the lowered cost of debt that green bonds can potentially offer (Kapraun et al., 2021).

Nevertheless, we cannot attribute this surge in green bond issuances solely to financial motives. Although the positive impact they have on the cost of debt is certainly appealing, it is not significant enough to fully offset the additional costs associated with issuing a green bond as compared to conventional bonds (Azhgaliyeva, 2021). Therefore, the decision to issue green bonds is not solely driven by financial incentives but is a complex interplay of a multitude of factors, including specific firm characteristics. State ownership, for instance, is one such major influencing factor (Bancel & Glavas, 2020). This implies that governments can exert influence over firms in which they hold significant stakes to further their climate-related political agendas. Research also finds that board gender diversity combined with debt maturity structure affect the decision to issue green bonds (Cicchiello et al., 2022).

Green bonds also serve as a powerful signalling tool, offering a way for companies to express their commitment to sustainable practices (Sangiorgi & Schopohl, 2021). The issuance of green bonds can be viewed as an element of a company's broader Corporate Social Responsibility (CSR), aligning with societal demands for environmentally friendly business practices (Bancel et al., 2023). It illustrates a proactive approach towards addressing environmental issues, thereby portraying a company as a responsible entity cognizant of its wider societal impacts.

Regulatory frameworks and institutional arrangements also play a key role in shaping the green bond market's dynamics. The Paris Agreement of 2015, which sought to "make finance flows consistent with a pathway towards low greenhouse gas emissions and climate-resilient development", prompted a discernible rise in the issuance of green bonds. It catalysed a shift towards an environmentally conscious financial paradigm, positioning green bonds as an instrumental tool in the fight against climate change. Further regulations, such as the French "Loi de Transition Energétique pour la Croissance Verte" and the EU Non-Financial Reporting Directive, added another dimension of responsibility for firms in terms of environmental disclosures. These regulations serve to increase transparency in corporate operations, thereby empowering authorities, investors, and the public with essential information that can guide their decision-making process. Transparency serves as a valuable tool for differentiating between firms based on their environmental performance, thereby incentivising companies to adopt more sustainable practices (Jankovic et al., 2022).

The issuance of green bonds is not an isolated decision but rather is intertwined with a complex web of institutional factors. Prior research posits that the institutional context significantly impacts how firms engage in CSR (Di Giuli & Kostovetsky, 2014; Kim et al., 2013; Yin & Zhang, 2012). Boubakri et al. (2019) highlight that in weaker institutional settings, the link between CSR and state ownership is more pronounced. This finding underscores a unique mechanism through which state ownership can direct firms towards more sustainable practices by endorsing and promoting the issuance of green bonds. A significant hurdle to green bond issuance is the increased cost associated with issuing green bonds compared to traditional bonds. Despite their countless benefits, this higher cost could deter firms from choosing green bonds. Despite such obstacles, the global drive towards sustainability, amplified by international commitments like the Paris Agreement and increased public demand for strict climate change regulations, has accelerated the adoption of green bonds.

As international institutions are increasingly pushing for more environmental regulation, green bonds emerge as a demonstration of the way financial instruments can contribute to the global fight against climate change. Public opinion and societal sentiment also have a considerable influence on the green bond market. Sustainability, being a critical concern for many, has resulted in growing demand for green investments. A survey in the U.S. found that 72% of Americans consider global warming to be personally important (Leiserowitz et al., 2019), while a study reported similar sentiments in Europe (McCright et al., 2015).

The risk of greenwashing accusations may influence firms' decisions to issue green bonds. Falchi et al. (2022) introduce the concept of "greenhushing", where companies under-communicate their sustainability efforts. This raises the question of whether some firms with green projects may avoid issuing green bonds to prevent scrutiny or accusations of greenwashing. Conversely, Flammer (2021) finds that green bond issuance positively impacts firms' environmental performance and serves as a credible signal of commitment to sustainability.

Considering the prevailing environmental challenges, green bonds have emerged as a key financial instrument aiding the transition to a low-carbon economy, despite a slight decrease in issuance in 2022. Influenced by a group of factors ranging from institutional settings, state ownership, CSR, to the reduced cost of debt they potentially offer, the decision to issue green bonds signals a shift towards sustainable practices and a proactive stance on environmental issues. As well as being a significant financial tool, green bonds serve as an indicator of corporate sustainability commitment, thereby contributing to the company's reputation and public perception. In addition, regulatory frameworks like the Paris Agreement and societal sentiment towards environmental consciousness play main roles in shaping the dynamics of the green bond market. Although the higher costs associated with issuing green bonds present a deterrent, the growing demand for sustainable investments and the worldwide drive towards environmental protection underscore the continued relevance and potential of green bonds as a tool for climate change mitigation.

Recent research has identified several factors influencing green bond issuance in Europe. Cicchiello et al. (2022) find that firm size, profitability, and environmental performance are positively associated with green bond issuance. Jankovic et al. (2022) highlight the importance of transparency in government green bond markets. Sangiorgi and Schopohl (2021) reveal that institutional investors are motivated by both financial and non-financial factors when investing in green bonds.

While understanding the motivations for issuing green bonds is important, it is equally important to examine the factors driving the overall growth of the EU green bond market. The next section explores the key elements contributing to this expansion.

What Explains the Expansion of the EU Green Bond Market?

European sustainabilism is a powerful catalyst driving change across a myriad of sectors (Hoerber & Weber, 2021), not least within the financial industry. It manifests in the form of robust policies, ground-breaking strategies, and evolving mindsets, all converging towards the goal of a more sustainable and resilient continent. Within this paradigm, green bonds have emerged as an innovative financial instrument that offers a practical way to fund projects with environmental benefits. Serving as a bridge between investors and projects with environmental benefits, green bonds present a viable solution to the conundrum of how to meet the growing demand for sustainable investment opportunities while contributing to the fight against climate change.

This revolutionary shift towards sustainable finance has been particularly evident within the EU, a leading figure in the global push towards sustainable development. The EU's stance on sustainability (Hoerber & Weber, 2021) has been instrumental in setting the stage for the proliferation of the green bond market. Rooted in ambitious policies such as the EU Green Deal, the European Commission's Sustainable Finance Action Plan, and the Paris Agreement, the

EU's commitment to a sustainable future has incentivised both public and private sectors to reassess their strategies and pivot towards greener practices. This convergence of policy and market demand has played a key role in the explosive growth of the EU green bond market. By placing a spotlight on the importance of green bonds, the EU is promoting a more sustainable, climate-resilient future, while simultaneously offering compelling opportunities for investors looking for both financial return and positive environmental impact.

The escalating growth of the EU green bond market is significantly influenced by several factors. These include advancements in green bond regulations, the institutional context of their issuance, the EU's unwavering commitment to the Paris Agreement, the recently introduced Taxonomy Regulations and the ambitious objectives of the overarching EU Green Deal. Each of these elements has been instrumental in fuelling the rise of the EU green bond market.

Green bond regulations have undergone a progressive evolution on a global scale. This progression has seen the developed world in particular take considerable strides in developing and implementing robust green bond regulations. This development of regulatory frameworks has been facilitated by a conspicuous shift in public sentiment, which has increasingly moved towards favouring more stringent climate-conscious policies and investment practices. In the face of a potential wave of stricter regulations, companies are now being legally incentivised to broadcast their good climate objectives more overtly. Simultaneously, investors are increasingly leaning towards low-carbon investments such as green bonds as a proactive strategy to mitigate the risk associated with harsher future regulations (Glavas, 2020).

Furthermore, institutional factors wield significant influence over the issuance of green bonds. Research conducted in recent years has shone a light on a strong correlation existing between CSR, state ownership, institutional weakness, and the issuance of green bonds (Glavas & Bancel, 2023). CSR has been recognised as a potent driver of sustainability efforts as businesses endeavour to harmonise their operations with environmentally friendly practices and societal expectations. The concepts of state ownership and institutional weakness have also played a substantial role in promoting the adoption of green bonds, driven by an emphasis on accountability and transparency.

The Paris Agreement, with its binding legal obligations, has been a significant catalyst for the rapid expansion of the green bond market, particularly in regions with robust regulatory frameworks such as the EU. This international consensus has pushed governments and investors towards sustainable practices, underpinning the need for stringent green bond standards to prevent greenwashing and ensure genuine environmental impact.

The EU's climate strategy has carbon neutrality by 2050 at its heart. This ambitious goal, aligned with the Paris Agreement, has cast capital markets in a key role. They are effectively the conduit through which the necessary funds to reach these targets can be channelled. The EU estimates that reaching these ambitious climate targets would necessitate an additional investment of around

€480 billion annually over the coming decade. This figure underscores the urgent need for significant enhancements to existing capital market structures and standards, as well as efforts to promote the comparability and standardisation of sustainable finance products (Born et al., 2021).

The EU's Taxonomy Regulation, introduced in June 2020, further strengthens the regulatory framework for green bonds. It provides a comprehensive classification system for environmentally sustainable economic activities. It is aimed at providing a universally accepted language to identify the degree to which an economic activity is environmentally sustainable, thereby resolving any ambiguity and fostering transparency for all market participants.

European sustainability reporting developments have been driven by a combination of legal and market-based factors, including the EU's Sustainable Finance Action Plan and the increasing demand for sustainability information from investors and other stakeholders. The CSRD is a legislative proposal made by the EU Commission that aims to establish a common set of rules for sustainability reporting across the EU and is expected to have significant implications for corporate practice and reporting standards, as it would require companies to report on a wide range of sustainability issues using a double materiality approach (Baumüller & Sopp, 2022).

The double materiality approach requires companies to report not only on the impact of sustainability issues on their own operations but also on the impact of their operations on society and the environment. The directive is expected to have significant implications for corporate practice and reporting standards. Companies will need to collect and report on a much broader range of sustainability information, which will require significant investments in data collection and reporting systems. The directive will also require companies to report on a wider range of stakeholders, including employees, customers, and suppliers, which will require more extensive stakeholder engagement.

The proposed directive is also part of the EU's Sustainable Finance Action Plan, which aims to mobilise private capital towards sustainable investments and to ensure that the financial system contributes to sustainable development. The directive is expected to increase transparency and comparability of sustainability information, which will enable investors and other stakeholders to make more informed decisions. Overall, the CSRD is a significant step towards a more sustainable European economy, but it also presents European companies with several challenges and forces them to face a new and considerably more demanding reporting environment.

On February 28, 2023, in a significant leap forward, the Council of the European Union and the European Parliament announced a provisional agreement for the creation of the EuGB. This move signals the development of a structured system for green bonds, fostering clarity, and ensuring credibility for investors. The proposed EuGB, although voluntary, mandates issuers to provide extensive disclosures about how the proceeds from bonds will be used. Additionally, they must demonstrate how these investments align with the company's overall transition plans towards more sustainable practices.

The creation of the EuGB is integrated into the broader European Green Deal, an encompassing strategy that aims to transform Europe into a sustainable and carbon-neutral continent by 2050. The Deal proposes a strict verification process for green bonds, which could potentially incentivise issuers to prioritise these instruments as their primary tool for sustainable financing. The EuGB has also spurred debate around its necessity and effectiveness. Critics argue it may harm the green bond market, citing insufficient private enforcement mechanisms and disruption to competition between various bond standards (Pyka, 2023).

Proposed modifications to the EuGB from the literature include (Pyka, 2023):

- Making the EuGB exclusive for all green bond issuances in the EU.
- Requiring private standards to align with the EuGB and be subject to external review and supervisory authority.
- Introducing effective private enforcement mechanisms for bond obligations and permitting additional private law obligations.
- Extending the standard's application to social and sustainable bonds.

These modifications could potentially address market deficiencies and foster a balanced public-private dynamic in the market.

It is also noteworthy that greenwashing remains a prominent concern within this industry. Even with the issuance of the EU's Directive on substantiation and communication of explicit environmental claims, commonly referred to as the "Green Claims Directive," its scope explicitly excludes financial products. As explicitly stated in paragraph 10 of the Directive, this legislative instrument does not apply to sustainability information involving messages or representations that may be either mandatory or voluntary pursuant to the Union or national rules for financial services. Such services include those related to banking, credit, insurance and re-insurance, occupational or personal pensions, securities, investment funds, investment firms, payment, portfolio management, and investment advice, as well as related settlement, clearing, advisory, intermediation, and other auxiliary financial services, including standards or certification schemes relating to such financial services. This exclusion raises questions about how best to ensure transparency and credibility in green claims made within the financial sector, which, at present, remains outside the purview of the Directive's regulatory measures. Additionally, to legal sanctions some solutions may include the use of a ratchet effect which corresponds to pushing some green bond issuers to fulfil their green commitments when they fail to do so (Glavas et al., 2023). One other behaviour that is underexplored in the literature is the importance of some bond issuers not using green bonds while they could, this behaviour can be assimilated to greenhushing (Falchi et al., 2022).

While the Green Claims Directive does not currently apply to green bonds, its principles could inform future regulations in this market. Extending similar transparency and accountability measures to green bonds could enhance their credibility and investor confidence. For instance, requiring standardised disclosure of the environmental impact of green bond-funded projects could help combat greenwashing concerns and provide investors with more reliable information for decision-making.

The proposed EU Directive on Green Claims, while not directly targeting financial products, raises questions about the need for similar measures in the green bond market. The directive's aim of "Ending greenwashing for Zero Pollution" could potentially be extended to green bonds, addressing concerns about the credibility of environmental claims in sustainable finance.

Despite these potential challenges, the rapid growth of the EU green bond market is indicative of a significant shift in investment behaviour towards aligning financial practices with long-term environmental sustainability objectives. This seismic shift not only represents a considerable stride towards a sustainable future but also exemplifies the important role the financial sector plays in global climate action.

Having examined the definition, motivations, and growth factors of the EU green bond market, we can now draw together these threads to understand the broader implications and prospects of this financial instrument.

Conclusion

The concept of green bonds represents a significant paradigm shift in the global and EU financial framework. The market expansion of green bonds can be primarily attributed to factors such as regulatory developments, institutional arrangements, climate commitments like the Paris Agreement, CSR strategies, state ownership, and shifting public sentiment towards sustainability. The EU green bond market, in particular, has been propelled by the EU's resolute dedication to the Paris Agreement, advancements in green bond regulations, the recent implementation of the Taxonomy Regulation, and the ambitious goals of the EU Green Deal. CSRD and the institution of the EuGB also play important roles in promoting transparency, accountability, and standardisation in the market.

This analysis, framed within the European environmental conscience methodology, has demonstrated how green bonds effectively connect energy and environmental policies. The innovative nature of green bonds as a sustainable finance solution is evident in their rapid market growth and evolving regulatory frameworks. The European Commission's expertise and leadership in developing standards like the EU Green Bond Standard and the Taxonomy Regulation underscore the EU's role in shaping the global green bond market. Furthermore, the increasing demand for green bonds reflects a growing grassroots support for sustainable finance, aligning with the core tenets of the European environmental conscience framework.

Despite the several benefits of green bonds, certain challenges persist. Increased costs associated with issuing green bonds, compared to traditional bonds, could deter firms. Regulatory debates, lack of private enforcement mechanisms, and potential limitations of standards present additional hurdles.

Nevertheless, the rapid expansion of the green bond market, despite these challenges, represents a significant stride towards aligning financial practices with long-term environmental sustainability objectives. This development highlights the fundamental role the financial sector can play in global climate action.

Future research could explore several key areas: the potential of green bond securitisation to increase Small and Medium-sized Enterprises' access to sustainable finance, the relationship between green bonds and commodities in the European context, and the long-term impact of the EuGB on market growth and credibility (Naeem et al., 2021). Additionally, investigating the effectiveness of green bonds in achieving environmental outcomes and their role in companies' broader sustainability strategies could provide valuable insights for policymakers and investors alike.

Bibliography

Azhgaliyeva, D. (2021). Green Islamic bonds. *Asian Development Outlook*.

Bancel, F., & Glavas, D. (2020). Why do firms issue green bonds? *Bankers, Markets & Investors*, *160*(1), 51–56.

Bancel, F., Glavas, D., & Karolyi, G. A. (2023). Do ESG factors influence firm valuation? Evidence from the field. *SSRN Electronic Journal*. https://doi.org/10.2139/ssrn.4365196

Baumüller, J., & Sopp, K. (2022). Double materiality and the shift from non-financial to European sustainability reporting: Review, outlook and implications. *Journal of Applied Accounting Research*, *23*(1), 8–28.

Born, A., Giuzio, M., Lambert, C., Salakhova, D., Schölermann, H., & Tamburrini, F. (2021). Towards a green capital market union: Developing sustainable, integrated and resilient European capital markets. *European Central Bank*. Verfügbar, 12.

Boubakri, N., Guedhami, O., Kwok, C. C. Y., & Wang, H. (Helen). (2019). Is privatization a socially responsible reform? *Journal of Corporate Finance*, *56*, 129–151. https://doi.org/10.1016/j.jcorpfin.2018.12.005

Cicchiello, A. F., Cotugno, M., Monferrà, S., & Perdichizzi, S. (2022). Which are the factors influencing green bonds issuance? Evidence from the European bonds market. *Finance Research Letters*, *50*, 103190. https://doi.org/10.1016/j.frl.2022.103190

Climate Bond Initiative (2022). Sustainable debt market summary. https://www.climatebonds.net/resources/reports/global-state-market-report-2022

Di Giuli, A., & Kostovetsky, L. (2014). Are red or blue companies more likely to go green? Politics and corporate social responsibility. *Journal of financial economics*, *111*(1), 158–180.

Falchi, A., Grolleau, G., & Mzoughi, N. (2022). Why companies might under-communicate their efforts for sustainable development and what can be done? *Business Strategy and the Environment*, *31*(5), 1938–1946.

Flammer, C. (2021). Corporate green bonds. *Journal of Financial Economics*, *142*(2), 499–516. https://doi.org/10.1016/j.jfineco.2021.01.010

Freeburn, L., & Ramsay, I. (2020). Green bonds: Legal and policy issues. *Social Science Research Network*. https://doi.org/10.2139/ssrn.3715969

Glavas, D. (2020). Green regulation and stock price reaction to green bond issuance. *Finance*, *41*(1), 7. https://doi.org/10.3917/fina.411.0007

Glavas, D. (2022). Do green bond issuers suffer from financial constraints? *Applied Economics Letters*, 1–4. https://doi.org/10.1080/13504851.2022.2083559

Glavas, D., & Bancel, F. (2023). Does state ownership impact green bond issuance? International evidence. *Finance*, I20-LII. https://doi.org/10.3917/fina.pr.020

Glavas, D., Grolleau, G., & Mzoughi, N. (2023). Greening the greenwashers – How to push greenwashers towards more sustainable trajectories. *Journal of Cleaner Production*, 382, 135301. https://doi.org/10.1016/j.jclepro.2022.135301

Hoerber, T., & Weber, G. (2021). Introduction – The European environmental conscience in EU politics 1. In *The European environmental conscience in EU politics* (pp. 1–23). Routledge.

Hoinaru, R., Benson, C., Stănilă, G. O., Dobre, F., & Buda, D. (2020). Green Bonds: Between economic incentives and eco-change. Proceedings of the International Conference on Business Excellence, 14(1), 236–245. https://doi.org/10.2478/picbe-2020-0022

ICMA (2018). And International Capital Market Association (ICMA) (2018). Green bond principles: Voluntary process guidelines for issuing green bonds.

Jankovic, I., Vasic, V., & Kovacevic, V. (2022). Does transparency matter? Evidence from panel analysis of the EU government green bonds. *Energy Economics*, 114, 106325.

Kapraun, J., Latino, C., Scheins, C., & Schlag, C. (2021). (In)-credibly green: Which bonds trade at a green bond premium? *Proceedings of Paris December 2019 Finance Meeting EUROFIDAI-ESSEC.*

Kim, C. H., Amaeshi, K., Harris, S., & Suh, C.-J. (2013). CSR and the national institutional context: The case of South Korea. *Journal of Business Research*, 66(12), 2581–2591.

Leiserowitz, A., Maibach, E. W., Rosenthal, S., Kotcher, J., Bergquist, P., Ballew, M., Goldberg, M., & Gustafson, A. (2019). *Climate change in the American mind: April 2019.* Yale University and George Mason University. Yale Program on Climate Change Communication.

Liang, X., & Gao, H. (2022). *11. Assessing the quality of green finance standards* (pp. 165–176).. Dans Open Book Publishers. https://doi.org/10.11647/obp.0328.11

Marcos, S., & Castrillo, M. J. (2022). Sustainable finance in Europe: The EU taxonomy and green bond standard. *Handbook of research on global aspects of sustainable finance in times of crises* (pp. 114–130). IGI Global.

McCright, A. M., Dunlap, R. E., & Marquart-Pyatt, S. T. (2015). Political ideology and views about climate change in the European Union. *Environmental Politics*, 25(2), 338–358. https://doi.org/10.1080/09644016.2015.1090371

Naeem, M. A., Adekoya, O. B., & Oliyide, J. A. (2021). Asymmetric spillovers between green bonds and commodities. *Journal of Cleaner Production*, *314*, 128100.

People's Bank of China. (2021). *Financial statistics report for 2021.* http://www.pbc.gov.cn/goutongjiaoliu/113456/113469/4342400/2021091617180089879.pdf

Pyka, M. (2023). The EU green bond standard: A plausible response to the deficiencies of the EU Green Bond Market? *European Business Organization Law Review.* https://doi.org/10.1007/s40804-023-00278-2

Sangiorgi, I., & Schopohl, L. (2021). Why do institutional investors buy green bonds: Evidence from a survey of European asset managers. *International Review of Financial Analysis*, *75*, 101738.

Yin, J., & Zhang, Y. (2012). Institutional dynamics and corporate social responsibility (CSR) in an emerging country context: Evidence from China. *Journal of Business Ethics*, *111*, 301–316.

6 Macroeconomic Impact of a Green Finance Shock

Gaël Callonnec, Mathieu Garnero, and Noam Léandri

Introduction

The European action plan on sustainable finance has been presented in 2018 and linked to the EU green deal in 2019. This holistic action plan integrates environmental conscience into the regulation of the economy and finance and is part of a drive to embed environmental issues into EU policies (Hörber & Weber, 2022). It raised awareness of the green financing issue which figures prominently in the Paris climate agreement (article 2.1.c) and remains in 2024 a decisive parameter in a context of public budget constraints. The road to carbon neutrality in Europe will require 1,000 billion euros of investments by 2030 according to the European Commission and all appropriate levers must be mobilised.

The financing of the economy and more specifically the banking sector has long been considered absent from the computable general equilibrium models used by macroeconomists, including within institutions such as central banks or the IMF. The credit crunch of 2008, the reinforcement of central bank intervention via quantitative easing policies has highlighted the importance of dealing more specifically with this issue by not considering the banking sector as a neutral intermediary between the company and investors.

ThreeME is a multi-sector macroeconomic model for evaluating energy and environmental policies, developed in this direction with the direct collaboration of the OFCE (French Observatory of Economic Conjunctures) and in consultation with the Ministry of Ecological Transition. It is used to estimate the macroeconomic effects of energy transition scenarios and ecological tax measures. The research work described in this paper focuses on enriching this model to better integrate the issue of financing the economy and "green" activities. ThreeME is a neo-Keynesian model. Unlike Walrasian neoclassical models, investment is not financed by savings. Profitable investments are assumed to be financed by the supply of bank credit, in other words, by money creation. This assumption leads to an underestimation of the crowding-out effect between investments and an overestimation of the investment multiplier. On the other hand, neoclassical models, by assuming a loop between savings and investment, neglect money creation and thus underestimate the capacity to achieve a low-carbon transition and its economic benefits. Like many general

DOI: 10.4324/9781032656359-9

equilibrium models, ThreeME's credit supply is not described in detail, in particular the banking sector is assumed to be a neutral intermediary of the central bank's monetary policy. This assumption was notably challenged by the banking crisis of 2008/2011. To improve the realism of the model, we have chosen to introduce a financial block composed of two parts:

1 A bond market linking the financing needs in the form of securities of companies and governments with the demand of households and central banks.
2 A banking sector providing bank credit to businesses and governments, the characteristics of which are determined by central bank policy.

The establishment of a financial block responds first and foremost to the need to consider certain rigidities related to the financing of the economy. By representing the role of savings in the financing of the economy, we generate a crowding-out effect specific to neoclassical models. This can have effects on the volume of credit and on prices, notably with a reinforcement of the rise in interest rates following an expansive fiscal policy by the state, leading to a less favourable multiplier effect (Ianc & Turcu, 2020). Beyond these new properties, the finer representation of the actors and the instrument of financing the economy allows for an analysis of new types of ecological transition scenarios, notably based on supply-side policies.

The establishment of a financial block should allow for an a priori evaluation of this type of policy in the ThreeME model, making it possible to distinguish the actors and instruments involved in these scenarios. These include:

- The central bank, which can intervene by buying public or private securities on its balance sheet ("Green Quantitative Easing");
- Private banks, which may be compelled by bank regulation to invest in certain sectors;
- Households, whose savings flow could be directed more towards certain sectors through the implementation of labels on market instruments (e.g. green bonds).

The paper reviews the main elements of the ThreeME model and describes the additions that were proposed during the internship to account for a more realistic financing model of the economy. It also presents the results obtained for different types of economic scenarios in this new modelling framework. The first part provides a brief description of the ThreeME model and estimates the additional investment needed to achieve a carbon neural target by 2050. The second part, more detailed, presents the modelling elements that have been added in ThreeME. We provide the theoretical justifications and bibliographic references used, as well as the choices made for the calibration. The last part consists of the exploitation of the results of the augmented model of the module on the financing of the economy. This work focuses first on the evolution of the properties of the macroeconomic model, and then on the possibilities offered by this extension of the model, with the addition of financing scenarios. In conclusion, we will come back to what this work tells us about the role of finance in the ecological transition.

Financing the Transition

ADEME used the ThreeME model to estimate the effects on growth, employment, income and the trade balance for the French economy (Callonec et al., 2013). Our study of various transition scenarios clearly shows that a decoupling between gross domestic product (GDP) and greenhouse gas emissions is quite possible. Especially thanks to the level of investment in green activities that are estimated bellow.

The ThreeME Model

ThreeME is a macroeconomic model developed by ADEME and OFCE since 2008. ThreeME is one of the main French general equilibrium models for linking the economy, the environment and energy needs. In this second part, we present a general description of the model, as well as a reminder of some of the economic properties of this neo-Keynesian model and some of its expected effects. It describes the French economy in 37 sectors, including 17 energy sectors, and integrates technical-economic aspects. The model has recursive dynamics and adaptive expectations. Energy consumption depends on the evolution of the stock of housing, vehicles, capital goods and their characteristics. Its main objective is to evaluate the medium- and long-term impact of the environmental and energy policy measured on the economy at macroeconomic and sectoral levels.

ThreeME has two key features:

1 First, it has the main characteristics of neo-Keynesian models. It assumes a slow adjustment of effective quantities and prices to their notional, i.e. optimal, level. The money supply is endogenous, a Taylor rule allows for the adjustment of rates to the macroeconomic context (inflation, growth, unemployment), and a Phillips curve or "wage setting" curve allows for the dynamics of employment in wages. Compared to classical multi-sectoral computable general equilibrium models, this has the advantage of allowing for the existence of sub-optimal equilibria, with for example the presence of involuntary unemployment.

2 The ThreeME model is hybrid: in addition to the macroeconomic loop, a bottom-up approach uses the description of the building and vehicle stock with its own renewal dynamics. Energy demand is derived from the evolution of this stock. This technical-economic modelling translates the possible energy substitutions, energy demand behaviours at the boundaries. The production and consumption structures are represented by a generalised CES utility function that allows the elasticity of substitution to differ between each pair of inputs or goods.

The ThreeME model can be used in the following cases:

- The study of the impact of a carbon tax;
- The analysis of subsidy policies for green investments in the building, automobile or public transport sectors;
- Measuring the impact of transitions in energy sectors.

Carbon Neutral Scenarios Designed by ADEME

ADEME has developed four energy transition scenarios aimed at achieving carbon neutrality by 2050 in France (ADEME, 2022). The trajectories differ depending on the levels of sobriety and expected technological progress as outlined by Figure 6.1, which shows a comparative analysis of four scenarios (S1 to S4). The levers of change are illustrated with respect to sobriety, efficiency, governance, and environmental impacts. Figure 6.2 provides a detailed breakdown of the characteristics and key elements of each scenario, highlighting the specific strategies and technologies associated with the four scenarios.

The first scenario, S1 "Frugal Generation", is the one that relies the most on sobriety. The transition is mostly based on the decline in the carbon-intensive goods production, a significant fall in the number of cars in circulation, by almost half compared to the trend, and contraction in new constructions.

The second scenario, S2 "territorial cooperation", which also mobilises sobriety, but in a more progressive way, is clearly more beneficial. The decline in industrial production and new constructions is more limited than in S1.

The third scenario, S3 "green technologies" is closer to the trend than the previous ones in terms of industrial production. The production of manufactured goods is almost equivalent, except for the number of cars in circulation.

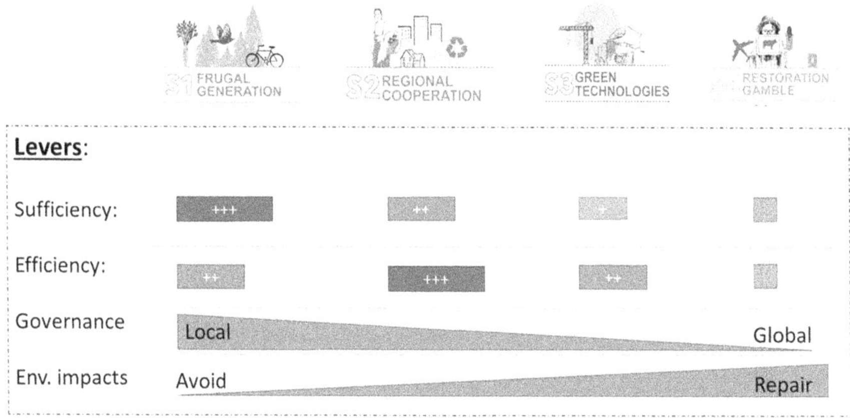

Figure 6.1 Comparative analysis of the four energy transition scenarios.

Source: ADEME (2022)

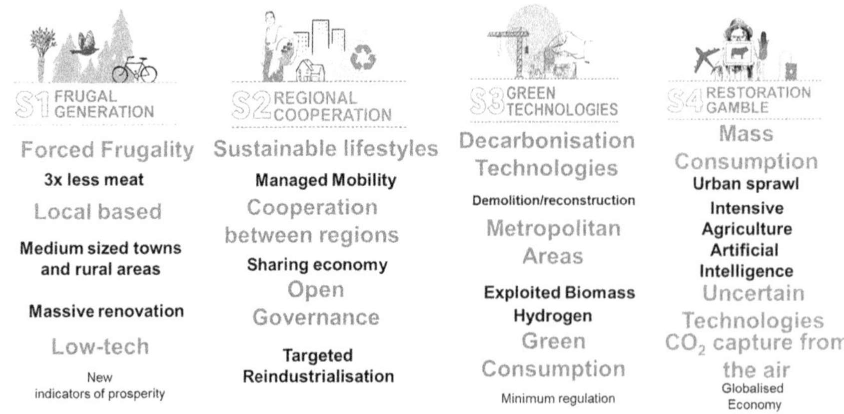

Figure 6.2 Characteristics and key elements of the four energy transition scenarios.
Source: ADEME (2022)

The fourth scenario, S4 "restorative bet" scenario excludes all kind of sobriety and relies entirely on technological progress to repair the impacts of economic activity. It is the most energy consuming. Industrial production is slightly higher than the trend. The number of new constructions and the number of vehicles in circulation are comparable. Achieving carbon neutrality presupposes additional decarbonisation investments in industry, which turn out to be less profitable than the previous ones, because the marginal cost of CO2 reduction is increasing.

Ultimately, none of the carbon neutral scenarios generates a recession relative to the current level of economic activity. The increase in activity linked to the transition comes from:

- A profitable substitution of imported fossil fuels (which prices are increasing) by renewables, produced locally.
- A substitution of manufactured products (which are carbon intensive and imported) by other local goods and services.
- Some profitable energy efficiency investments for households and businesses.

Level of Investment in Green Sectors

The transition involves increased investment in green sectors to the detriment of energy-intensive sectors. The balance is negative in S1 and S2 but positive in S3 and S4. These direct investments have a positive impact on the business. The effect is multiplier. A macroeconomic loop is therefore necessary to measure the overall evolution of the demand for capital.

After macroeconomic closure, the level of investment in scenarios S1 and S2 is lower than that of the trend, as shown in Figure 6.3 below.

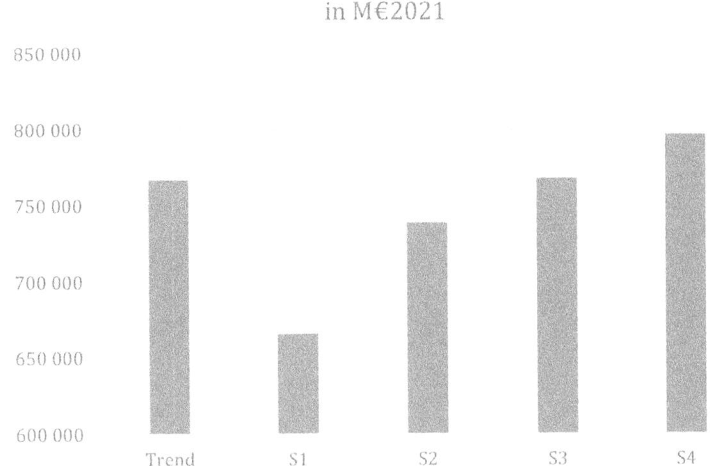

Figure 6.3 Level of investment of the four energy transition scenarios.

Source: ADEME (2022)

The growth rate of S2 is slightly higher than the trend, and yet the inflation is less important because the substitution of renewable energies for fossil fuels and the modification of the nature of the household consumption basket led to a decrease of its weighted average price. In these first two scenarios, it is very unlikely that the central bank will increase its key rates since the inflation rate is contained below the 2% mark and unemployment increases in S1 and remains practically stable in S2. A restrictive monetary policy is not to be feared. Bond yields are unlikely to rise as aggregate investment demand declines in both scenarios. The weighted average cost of capital for investors should therefore decrease. In short, if carbon neutrality is achieved through consistent sobriety efforts rather than technological progress, agents are unlikely to face funding constraints. It is different in scenarios S3 and S4 as we observe contrasting investment needs as shown by Figure 6.4.

In S1, investments decrease in industry because of the decline in demand in transport and building but also energy efficiency. The drop in industrial production has a negative impact on employment, which negatively affects the activity of the tertiary sector, which also limits its investments. Investments in renewable energies rise except for renewable heat.

In S2, the additional investment in industry and services amounted to nearly €26.3 billion on average per year over the period. Or 0.9% of current GDP. Investments in energy management alone amount to 48 billion on average per year in S2 against 35 in the trend.

In S3, there is an additional investment in industry and services of €18 billion on average per year, or 0.7% of current GDP. Green investments alone amount to 42.7 billion on average per year in S3 against 34.9 in the trend,

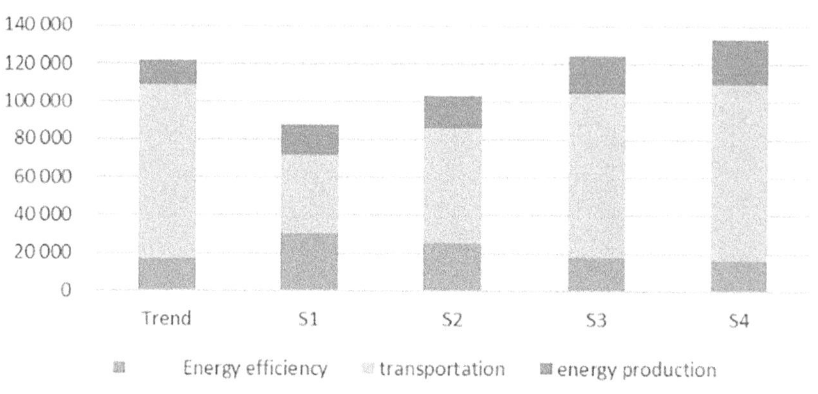

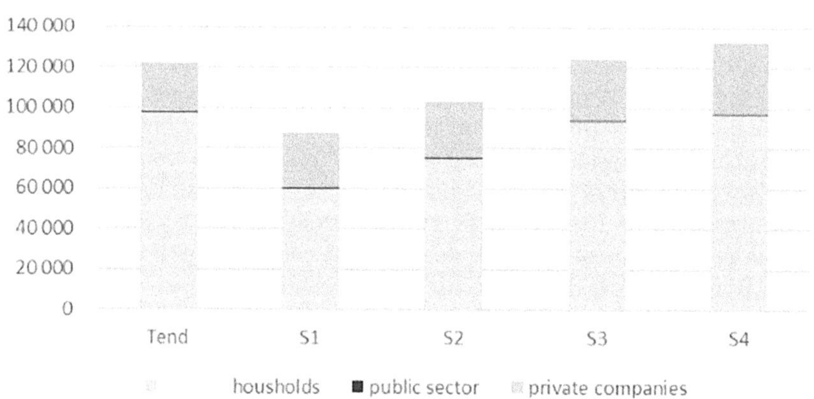

Figure 6.4 Contrasting investment needs for green transition.

Source: ADEME (2022)

i.e. 0.3% of current GDP. With the decrease in fuel-efficient vehicle purchases, the gap in green investments between the trend and S3 is now only €2.5 billion/year.

The S4 scenario is the most investment-intensive because:

- The estimated final energy demand is close to the trend (113 Mtp vs 126 in the trend).
- The number of vehicles in circulation and new constructions are close to the trend.
- The industrial production index is 103 in 2050 versus 102 in 2050 in the trend (base 100 in 2022).

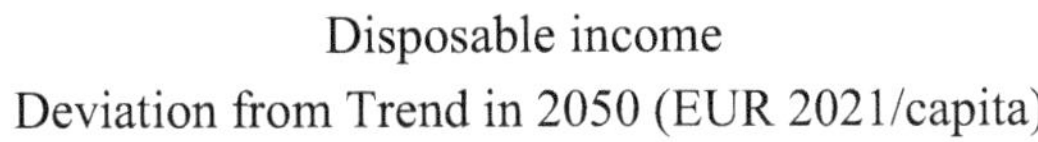

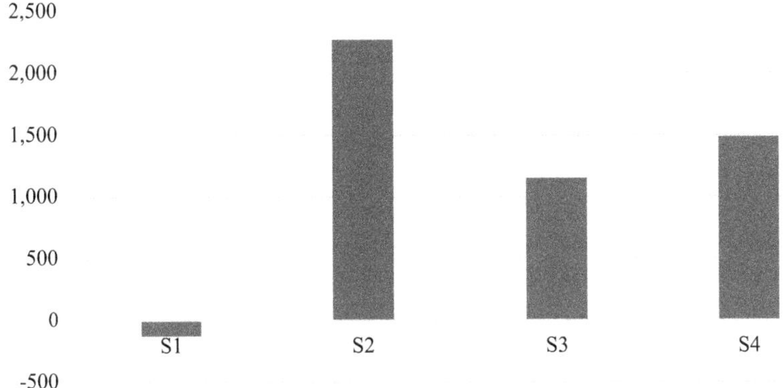

Figure 6.5 Profitable investment of the four-energy transition scenarios.

Source: ADEME (2022)

These expenses should not be considered as a cost, since they are lucrative thanks to the increase in the fuel prices as shown by Figure 6.5. These investments have a positive net present value (NPV): the sum of the cash flows is greater than the repayment of the loan (interest included). That is why the households' disposable income, after the debt repayment and the energy bill is rising in comparison to the trend. The rise of the private debt is then sustainable. It generates a strong value creation. Therefore, there is a decrease in the private debt/GDP ratio.

The Public debt/GDP ratio is also decreasing in the S2 S3 and S4 scenarios thanks to the growth of economic activities. The additional revenue collected by the State is greater than the increase in its expenditure. This demonstrates the benefit of a temporary relaxation of the Stability and Growth Pact (SGP) to finance the ecological transition and then benefit from the positive effect.

The Inclusion of Financing in Macroeconomic Models

To consider the financing of the economy in *ThreeME*, a module on the financing of the economy has been created, called the financial block. It considers two possible sources of financing: the first is provided by the banking channel via loans, i.e. a form of money creation by private banks. The second is provided by the securities market, in which two actors are involved: households, who invest their savings, and central banks, which purchase bonds as part of their monetary policy. The models used here must allow us to transcribe the evolution of the supply of financing according to the economic and market context. For the sake of robustness and consistency with the granularity of the rest of the model, we have introduced quite simple and aggregated models that have been used by other authors and institutions in similar modelling cases.

Objective of introducing a financial block

The addition of a financial block should allow for two important inputs:

1 Reinforce the realism of the model and its effects (investment multiplier, crowding out effect) in the framework of existing analyses.
2 Broaden the range of scenarios that can be implemented, opening the possibility of proactive green financing policies.

Furthermore, the addition of a financial block should allow for a more realistic representation of the two main means of financing the economy:

- The allocation of savings, which is limited in quantity and subject to a crowding-out effect linked to scarcity and the choice of investment;
- Money creation, implemented by private banks and the central bank.

We will look at the effects of this financial block to see if it results in a crowding-out effect at realistic properties. The crowding-out effect is generally the consequence of the expansion of the activity of one sector at the expense of another.
There are two types of eviction:

- Direct crowding out when the increase in public spending is instantly compensated by taxes, which in turn reduces private spending.
- Indirect (or financial) crowding out, which increases the cost of financial resources for companies through access to the bond market: higher rates following an increase in the supply of bonds.

If the direct effect seems to be of little relevance in our analysis (no full employment, no compensatory tax increase to finance the investment), we will focus on the indirect effects.

They are of two intrinsically linked natures:

- A volume effect: additional investments mobilise a share of savings to the detriment of other investments, which we will observe in the bond market,
- A price effect: increase in rates due to the drying up of the markets, observable on the bond and banking markets.

Modelling the banking sector

General Principle

The banking sector is represented in its component of bank credit supply to the private and public sectors. This supply of credit is characterised by two dimensions: the quantity of loans that can be granted and the interest rate charged. Like the other agents (government, households), the banking sector will be represented in aggregate as a single agent.

Research on the role of the banking sector in the 2008–2011 crisis has been the subject of many recent publications, notably by central banks, supervisors and supranational bodies. The main objective is to understand how the banking sector plays its role of monetary policy intermediation (Blundell-Wignall & Roulet, 2013), the other determinants of credit policy by banks (Gambacorta & Marques-Ibanez, 2011) and how the system is likely to respond to increased regulatory capital requirements (Roger & Vlček, 2011). As regulated institutions, banks have constraints on the amount of financing they can provide, including capital, expected loss provisions and liquidity.

Without going as far as stochastic modelling of the sector, as central banks may wish to do in their Dynamic Stochastic General Equilibrium (DSGE) models (Bernanke et al., 1999), we have chosen to use the main elements of an approach adapted to Computable General Equilibrium Model (CGE) designed jointly by the European Central Bank (ECB) and the European national central banks (Ehrmann et al., 2001). This work was carried out within the framework of the research network "Monetary Transmission Network" published by Cambridge University Press (Angeloni et al., 2003). The model was then used by the Bank for International Settlements (BIS) and the OECD in their macroeconomic analyses.

Business Model

Here we adopt the model and notations from the paper by Ehrmann et al., 2001. The model is based on the work of Stein (1995), who uses the stylised asset/liability balance sheet of a bank. Starting from the main elements of this balance sheet, it is possible to construct a profit function, and then to find the quantity of loans that will maximise this function.

This corresponds to the following equation:

$$L_i + S_i = D_i + B_i + C_i$$

Each component is modelled with a specific equation:

The loan demand is a function of the proposed rate $r_{L,i}$, aggregate output y and price levels p The bank's capital is evaluated as a percentage of the loans granted (principle of leverage ratio existing in the regulations).

The amount invested in securities is assumed to be proportional to the deposits (to maintain liquidity).

Deposits are non-interest bearing and are therefore negatively correlated with the risk-free rate r_S .

Finally, bank financing on the markets depends on the "health" of the banks as perceived by the market x_i which is a function of various factors such as size, liquidity or capitalisation of banks

By noting φ_i the operating costs of bank i, a simplified evaluation of the bank's profits is given by the equation:

$$\pi_i = L_i \times r_{L,i} + S_i \times r_S - B_i \times r_{B,i} - \varphi_i$$

By substituting the above equations into the profit function and then deriving to first order, the authors find the profit-maximising quantity of loans:

$$L_i = f\left(y, p, r_S, x_i, \varphi_i\right)$$

Econometric Model

The econometric estimation follows the loan supply equation presented above. It is estimated in first differences, considering a lag effect over j periods and an evaluation of cross effects.

$$\Delta \log\left(L_{it}\right) = a_i + \sum_{j=1}^{l} b_j \times \Delta \log\left(L_{it-j}\right) + \sum_{j=1}^{l} c_j \times \Delta r_{t-j} + \sum_{j=1}^{l} d_j \times \Delta \log\left(GDP_{t-j}\right)$$

$$+ \sum_{j=1}^{l} e_j \times infl_{t-j} + f \times x_{it-1} + \sum_{j=1}^{l} g_1 \times x_{it-1} \times \Delta r_{t-j}$$

$$+ \sum_{j=1}^{l} g_2 \times x_{it-1} \times \Delta \log\left(GDP_{it-j}\right) + \sum_{j=1}^{l} g_3 \times x_{it-1} \times infl_{t-j} + \varepsilon_{it}$$

The calibration is based on two data sources:

- A BankScope database provided by the rating agency Fitch which covers about 50% of the banks in the euro zone (about 3000) with annual data over some 4 years (1998–2001)
- A database generated by national supervisors (Eurosystem dataset) covering France, Germany, Italy and Spain with quarterly data over 4 years (1998–2001)

This econometric calibration in the case of France provides the long-run coefficients in Table 6.1.

In our model, we retain the following parameter values:

- For the effect of monetary policy: $c_j = -1.8$
- For the business cycle (GDP) effect: $d_j = 3$

Bond market modelling

General Principle

The second component of the financial block is credit financing in a bond market. According to our national accounts matrix, the supply of bonds is the primary source of financing for the government (91%) and for the private sector (68%). This financing is provided by households, who use it as an investment for their savings. The central bank, which can buy and place sovereign and non-sovereign bonds on its balance sheet, also holds a portion of this financing (see §2.3). Financing by the rest of the world is not modelled here.

At this stage, given its monetary policy announcements, it is assumed that the central bank, which pursues environmental objectives as secondary target, does not behave as an investor. It is assumed, in accordance with its announcements, that it acquires a fixed share of the bond debt issued, which ensures that it will renew its matured securities, regardless of developments in public or private bond rates.

Regarding households, the choice of investment of their wealth can be made on three assets:

- Monetary reserve with a risk-free rate R_M
- Public bonds paying R_{Opub}
- Private bonds paying R_{Opr}

Using standard microeconomic framework, we will assume that the investment choices made by households will depend on the relative levels of rates, which determines allocation.

Table 6.1 Econometric calibration for France

Models estimated with the following bank characteristics variables

	Size	Liquidity	Capitalization	Size, Liq. Capitalisation	Size liquidity
Monetary policy	−1.564** *0.765*	−2.131*** *0.736*	−1.823*** *0.701*	−1.969*** *0.566*	−2.221*** *0.697*
Ral GDP	3.239*** *0.578*	3.999*** *0.493*	3.788*** *0.503*	2.975*** *0.374*	2.523*** *0.470*
Prices	−2.850*** *0.742*	−4.173*** *0.692*	−3.701*** *0.689*	−3.678 *0.512*	−3.147*** *0.644*
Char1*MP	−0.458 *0.553*	4.030 *4.734*	3.547 *15.236*	−0.063 *0.218*	−0.184 *0.235*
Char2*MP				8.106*** *1.931*	7.070*** *2.010*
Char3*MP				2.304 *7.007*	
Char1*Real GDP	−0.262 *0.785*	−1.255 *7.508*	−16.48 *25.648*		
Char1*Prices	−0.070 *0.714*	−1.637 *6.143*	5.303 *24.352*		
Char1*Char2*MP					0.390 *1.228*
p-val Sargan	0.142	0.233	0.111	0.231	0.075
p-val	0.014 0.451	0.006 0.326	0.017 0.542	0.000 0.387	0.000 0.450
N. of banks, obs.	312 5327	312 5327	312 5327	312 5327	312 5327

Source: Ehrmann et al. (2001)

Microeconomic Model

We revisit the formalisation of Arrow's (1965) portfolio choice model, as presented in the subsequent work of Arrondel et al. (2007).

In Arrow's simplified model, an agent lives for a single period and has a stock of wealth w_0 that can be invested in two assets:

- A risk-free asset paying an interest rate of r
- A risky asset paying a rate $\tilde{r} = r + \tilde{z}$ where $\tilde{z}$ is a random variable representing the spread that remunerates the risky asset

The final utility of the agent is:

$$u\left(w + \alpha \tilde{z}\right)$$

where $w = w_0\left(1 + r\right)$ is the wealth obtained by investing in the risk-free asset and alpha is the amount invested in the risky asset.

The optimal risky investment α which maximises the utility expectation $argmax_\alpha \ E[u(w + \alpha \widetilde{z})]$ is given by limited expansion:

$$\alpha \approx \frac{E\left[\tilde{z}\right]}{\sigma_z^2 A_u\left(w\right)}$$

where $A_u\left(w\right) = \dfrac{-u''\left(w\right)}{u'\left(w\right)}$ is the Arrow-Pratt coefficient that represents the risk aversion of the agent with wealth w and utility function u.

We take the utility function proposed in the article which assumes constant relative risk aversion (CRRA)

$$u\left(C\right) = \frac{C^{1-\rho}}{1-\rho}$$

In the extensions proposed by the authors, they take into account the work of Drèze and Modigliani (1970). The authors consider a risky income-related component $\tilde{y}$ which affects household wealth.

$$u\left(w + \alpha \tilde{z} + \tilde{y}\right)$$

The optimal risky investment $\hat{\alpha}$ which maximises the utility expectation is given by Taylor expansion:

$$\hat{\alpha} \approx \frac{E\left[\tilde{z}\right]}{\sigma_z^2 A_v\left(w\right)}$$

Where the utility function $v(t) = E\left[u(t+\tilde{y})\right]$ has been modified to use a deterministic approach.

Calibration

As a benchmark, the work of Gollier (2002) deal with the question of risk aversion, they use the following parameters as reference:

- A central value for the Arrow-Pratt coefficient $A_u(w) = \dfrac{-u''(w)}{u'(w)} = 4$

- Volatility $\sigma = 0.16$ calibrated on stocks
- A risk-free rate $r = 1\%$

This corresponds to:

$$\frac{(1+r)}{\sigma^2 \times \rho} \approx 10$$

So that:

$$\Delta \propto = 10 \times \Delta E\left[\tilde{z}\right]$$

This means that a change of 1% in the expected rate of risky investment leads to an increase of 10% of the share of risky assets in the portfolios. Different hypotheses have been tested in an alternative way afterwards to see the sensitivity of the model to this parameter, especially a value of 5 instead 10.

Impact of a Green Financing Shock

The model thus augmented with a more precise financing module will have significantly different properties from the initial model, which we will analyse in the first part of our impact measurements. We will then be able to see the new applications allowed by this finer description of the financing offer, by using our model to evaluate the effect of voluntary policies of green financing shock. It will be interesting to see how a financing policy complements a simple budgetary policy and the long-term gains that this combination can bring.

Main Results

Multiplier Effect

We will now look at the macroeconomic effect on real GDP of a public fiscal policy shock of 1% of GDP. The response to this shock is called the investment multiplier; we expect a response value greater than x1.32. In our analysis, we

Table 6.2 Evolution of the investment multiplier according to the ThreeME and Mesange models

	1 year	2 years	3 years	5 years	10 years old	25 years old	LT
ThreeME - without financial block (FB)	1.36	1.37	1.31	1.10	0.77	0.53	0.46
ThreeME - with financial block (FB)	1.38	1.40	1.33	1.11	0.74	0.39	0.28
Mesange	1.31	1.38	1.22	0.93	0.57	0.30	0.22

Source: Authors

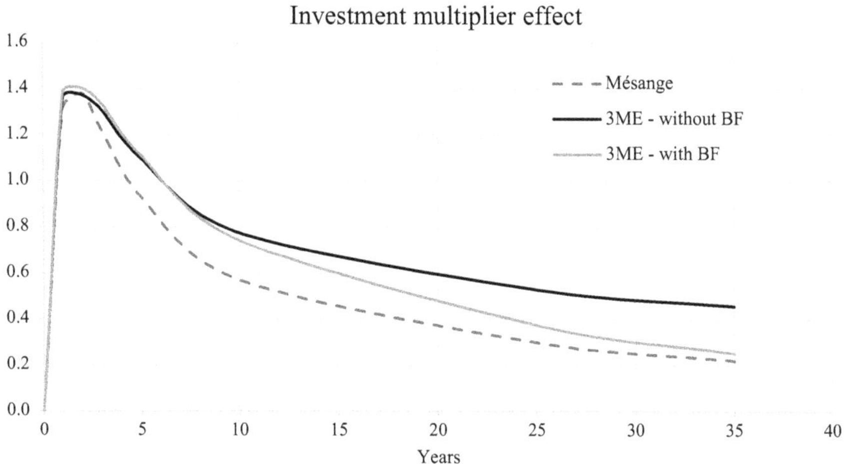

Figure 6.6 Investment multiplier effect.

Source: Authors

compare ThreeME with Mesange model which is the macroeconomic model used by the French Ministry of Economy and Finance (Table 6.1).

The results obtained by the three models are provided in Table 6.2 below.

The addition of the financial block (BF) in ThreeME results in an investment multiplier that is almost unchanged in the short term (from 1.37 to 1.40 at 2 years) as shown in Figure 6.6.

However, the effect is more noticeable thereafter, since in the long run (35 years), the persistence of the multiplier effect is x0.28 compared to x0.46. We note a greater proximity with the results of the variants of the Mesange model in the medium to long term (Bardajo et al., 2017).Year

Table 6.3 Average current interest rates following an investment shock (1% GDP)

Year	0	1 year	2 years	3 years	5 years	10 years	15 years
Without financial block (FB)	3.86%	6.56%	4.06%	3.83%	3.82%	4.01%	3.99%
With financial block (FB)	3.86%	6.89%	4.43%	4.15%	4.01%	4.19%	4.21%
Rate increase	0.00%	0.33%	0.37%	0.32%	0.19%	0.18%	0.22%

Source: Authors

Crowding Out

We then look at the crowding out effect between public and private investment following a public investment shock of 1% of GDP. Table 6.3 describes the average current interest rates following an investment shock. In the model without a financial block, the main channel for the crowding-out effect is through the feedback loop from investment to prices (inflation), which in turn leads to an upward adjustment of policy rates, according to the Taylor rule. In the model with a financial block, the drying up of the bond market through the increase in demand for public bonds should reinforce the rise in interest rates (price effect) and potentially reduce the amount of private investment (volume effect).

As expected, the crowding out effect on volumes is slightly more pronounced in the model with a financial block. However, two flexibilities introduced in the model limit the restriction on demand: households will increase their demand for bonds following the rise in interest rates (reduction in their monetary reserves) and private banks will increase their lending in response to the combined rise in demand and key rates.

The inflationary effect on rates is between +0.3% to +0.5%, when we compare the rate obtained by applying the Taylor rule, used in the model without financial blocks, and the weighted average of bank and bond rates on new loans.

The price effect of crowding out leads to an increase of about 7% in the interest burden on the private sector (about €10 billion per year in the first years).

Application to Financing Policy Shocks

General Principle

The scenarios usually implemented with ThreeME relate to public investments, subsidies for certain technologies and taxes on emissions. Here we will enrich this range of scenarios by analysing the effects of a policy of greening the supply side through financing facilities.

In the context of the sustainable finance action plan taken by the European Commission[1] in 2018 and the climate-related action plan of the ECB[2] released in 2021, several supply-side initiatives can be analysed:

- A green financing policy via a transfer of household savings: The European Commission's action plan aims at revealing savers' sustainability preferences.
- A green financing policy via private banks: The modification of banks' practices includes climate stress tests and a policy of transparency linked to the sustainable taxonomy.
- A green financing policy in the quantitative easing operations of the European Central Bank.

We have chosen to implement shocks compatible with the additional green financing needs evaluated by I4CE (10 to 30 billion euros per year for France, i.e. between 0.4 and 1.3% of the GDP/year). Working within the simplified framework of a 2-sector model, we cannot provide a complete analysis of these policies at this stage.

However, we will look at the effect of such policies on our simplified economy. Table 6.4 describes the potential effects.

We will test different possible configurations for funding:

1 An additional shock to the supply of financing in a sector
2 A substitution shock in the supply of financing between one sector and another

The impact of these policies is very different depending on whether they are additional or substitute financing. In the case of additional financing, the overall effect should be an increase in GHG emissions, except in the case of decoupling where investments would generate a decrease in emissions greater than the increase generated by the increase in GDP. If financing is substituted for other financing, i.e.

Table 6.4 Simplified tests of investment scenarios

Scenario	Description	Simplified test
Savings shock	Increase in savings invested in green sectors.	Increase in household savings in the private sector (+€10 billion/year).
Shock of financing by private banks	Trend increase in "green" loans (and possible trend decrease in "brown" loans).	Increase in private bank financing of the private sector (+€10 billion/year).
Central bank funding shock	Increase or greening by the central bank of the financing held on the balance sheet.	increase in the allocation of central bank bonds to the private sector (+€10 billion/year).

Source: Authors

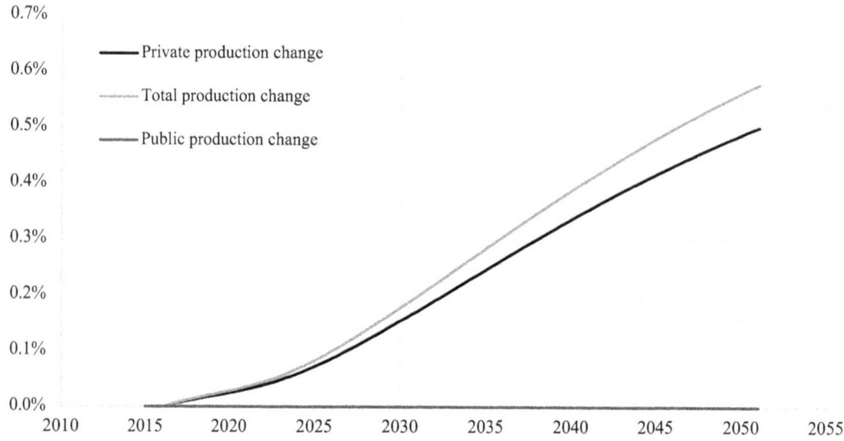

Figure 6.7 Evolution of production.

Source: Authors

if we simulate a transition policy marked by crowding out effects, we expect a greater decrease in GHG emissions with the acceleration of the renewal of the transport fleet and the renovation of the building stock, for example.

It will also be interesting to see the combined effect of a policy of stimulating financing with an investment policy. Implementing only financing policies will not necessarily provide the projects to be financed, without a concomitant stimulation of investments.

The savings shock scenario (+€10bn/year from 2017) corresponds to an additional allocation of household savings to bonds. Without an investment shock, this excess demand will not find a direct outlet and will be reflected in a lasting decline in bond interest rates, which will gradually stimulate investment.

Figures 6.7 and 6.8 show that in the long term (2060), the additional 10 billion in annual financing translates into 9 billion euros of private investment in volume terms, as well as a positive effect of 2.7 billion euros on public investment, also generated by the fall in bond rates. The main disadvantage that can be seen here is the very long time required for a financing policy alone to translate into effects on investment (more than 10 years for real effects).

We now look at the combined effect or otherwise of a public investment policy with a private sector financing policy.

This graph shows the favourable effect of a combination of financing and investment policies. Thus, the effect of the policy is sudden, and the investment multiplier is greater than 1 in the long run. Comparing the combined effect of the two policies with the sum of the individual effects, we see that the

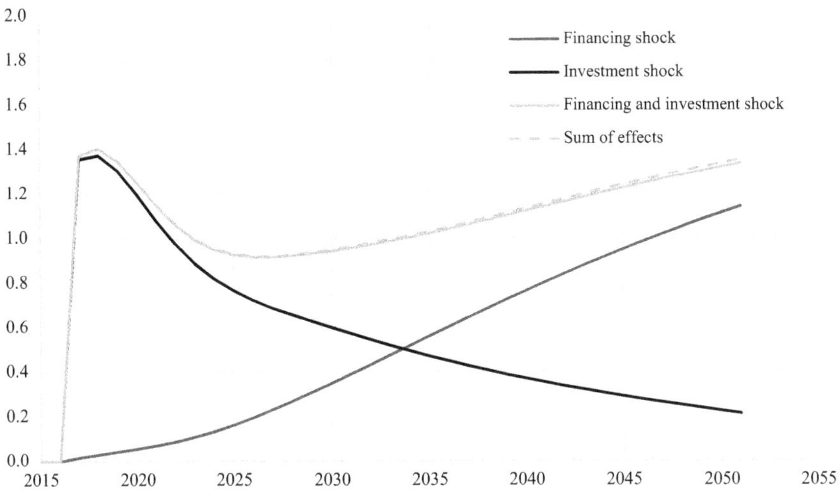

Figure 6.8 Impact on production expressed in investment multiplier.

Source: Authors

cross-term is almost zero (no crowding out effect between the two policies). The combination of the two policies thus produces beneficial effects while providing greater policy coherence in the short, medium and long term.

The analysis of the components of real GDP provides us with an additional view of the long-run quality of a financing shock as described in **Figure 6.9**.

A government spending shock stimulates investment in the short term but penalises it in the long term, just as it penalises the trade balance. GDP is then driven by government spending and household consumption. In the case of a financing shock, the breakdown of effects is strictly opposite, since it is essentially the trade balance and investment that explain the long-term growth surplus. In the case of a combined financing and investment shock, the imbalances of the government spending shock are slightly attenuated in the long run; in particular, a positive effect on investment is maintained.

Shock of financing by private banks and central banks provides quite similar effects on short- and long-term components of GDP, the differences are discussed in the following.

Comparison of the Three Financing Policies Shocks

In summary, we wanted to return to the comparative advantages of the three green financing policies.

In all three cases, the financing shock alone does not find its outlet in the short term, since the addition of €10bn of private sector financing does not

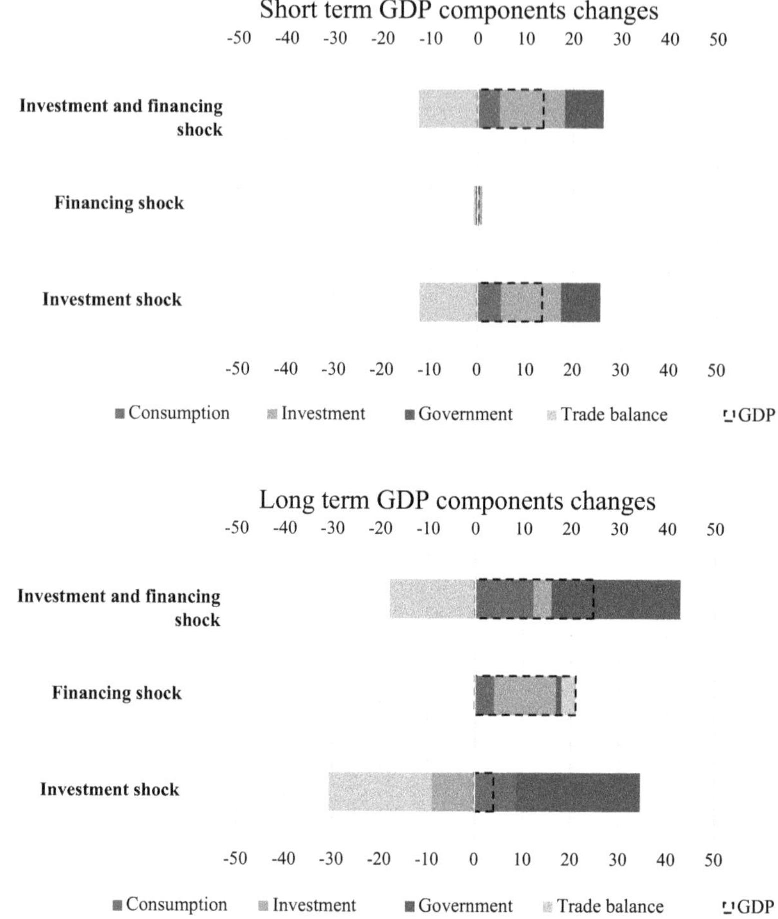

Figure 6.9 Short and long-term GDP components changes.

Source: Authors

create €10bn of investment as shown by Figures 6.9 and 6.10. The credit supply shock is thus reflected in the short term by a single effect on interest rates, which is relatively similar after two years: around −15bp at two years for public bonds and −30bp for private bonds, and 24bp on average (public and private bonds combined). It should be noted that an increase of €10bn in bond savings (+2.75% of annual invested savings) corresponds in our model to a theoretical decrease in rates of 27.5bps, if we simply consider the partial equilibrium on bond supply and demand given by the formula $d(\varphi_O) = \theta_O \times d(R_O - R_M)$ where $\theta_O = 10$. The result obtained is therefore very consistent with this relationship.

In the long run, the savings shock has a stronger effect on private bond rates: they remain permanently lower by about 35bps, while they rise slightly in the case of a shock to private bank or central bank financing.

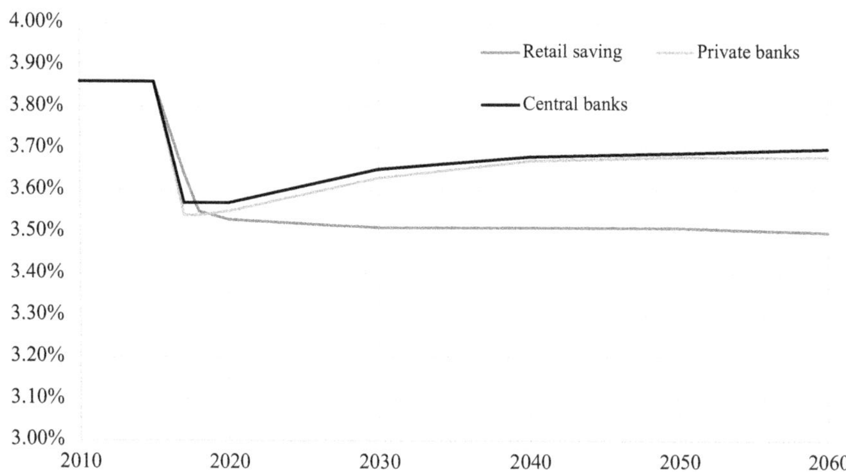

Figure 6.10 Effect of a financing shock of 10Md € on obligation rates.

Source: Authors

In the case of a savings shock, the fall in interest rates allows the public sector to reduce its debt, which ultimately results in a fall in the supply of securities that is greater than the fall in the maximum demand for securities linked to the loss of household wealth. In fact, in the long term, the disposable assets of households are doubly reduced by the fall in interest rates and by the fall in the volume of securities issued (deleveraging of the public sector). This excess demand for securities produces the additional drop-in long-term rates.

Conclusion

The design of a general equilibrium model with a financial block has allowed us to go one step further in the realism of the representation of financing in our macroeconomic modelling. The consideration of this issue is rather recent in this type of model (essentially since the 2000s) and the use of macro-econometric work conducted by central banks and international institutions, particularly following the 2008 crisis, provided the main references for our modelling.

The initial analyses conducted with our model confirm the general properties of the model without the financial block but make it possible to limit in the long run certain drifts associated with the sole representation of financing by money creation. In the end, in the context of an investment shock, the crowding-out effects introduced by the financial block are more related to price effects (higher interest rates) than to volume effects (investment restrictions).

Thus, in the first place, our approach can help calibrate a more realistic Taylor rule for this type of friction between credit supply and demand.

We have also used the flexibilities offered by our model to introduce three types of green financing shocks: by savings, by the private banking sector and by central banks. The analysis shows, first, that a simple financing policy has a very gradual effect over time, and that its effects only really materialise in the long term (beyond 10 or 20 years). From this point of view, the model indicates that stimulating savings provides the best results by reducing rates more durably. In fact, shocks to bank financing ultimately led to a reduction in household wealth through a dual effect of substitution of credit supply and a decline in financial returns. Moreover, we have been able to illustrate the benefits of a policy that combines an investment shock with a savings shock: it allows a positive response over different time horizons and thus avoids both the delays of a financing policy and the temporal crowding out of the effects of an investment policy.

This analysis could be followed by proposals for policy measures that could translate these scenarios into action.

On household savings:

- Greening regulated savings such as Livret Développement Durable et Solidaire, Livret A, PEL (€834 billion in outstandings at the end of 2021) (Banque de France, 2021).
- Stimulate, through information or taxation, the green or sustainable units of account in life insurance (€1677 billion in assets at the end of 2021).

This solution seems the simplest to implement in France, given the national policy levers and the very large stock of savings.

For private banks, the regulatory consensus is not currently very favourable to incentive policies (green supporting/brown penalising factors) via prudential capital requirements that would disrupt the supervision of financial risks. Moreover, as these are international or European regulations, the capacity to implement this type of policy seems limited today.

Regarding central banks, actions are already underway, particularly within the framework of the NGFS (Network for Greening the Financial System), a network of central banks aiming to voluntarily contribute to the implementation of the Paris Agreement. Given the amounts available to central banks (around 20% of GDP in bonds), the voluntary redirection of 1 to 1.5 points of GDP from these bonds to green bonds would be sufficient to meet the €60 billion financing gap identified by the I4CE panorama (2022).

Notes

1 "Action Plan: Financing Sustainable Growth", European Commission, 2018.
2 ECB action plan to include climate change considerations in its monetary policy strategy", 2021.

Bibliography

ADEME. (2022). *Transition(s) 2050. Decide now. Act 4 climate* [Executive summary]. ADEME.

Angeloni, I., Kashyap, A. K., Mojon, B., & Eurosystem Monetary Transmission Network (Eds.). (2003). *Monetary policy transmission in the Euro area: A study by the Eurosystem Monetary Transmission Network*. Cambridge University Press.

Arrondel, L., Calvo Pardo, H. F., & Oliver, X. (2007). *Temperant portfolio choice and background risk: Evidence from France* (Working Paper Paris School of Economics 2007–16). Paris School of Economics. https://papers.ssrn.com/sol3/papers.cfm?abstract_id=995519

Arrow, K. J. (1965). Aspects of the theory of risk bearing. *Yrjo Hahnsson Foundation*.

Bardajo, J., Campagne, B., Khder, M.-B., Lafféter, Q., Simon, O., Dufernez, A.-S., Elezaar, C., Leblanc, P., Masson, E., & Partouche, H. (2017). *The Mesange macroeconometric model: Re-estimation and innovations* (INSEE Working Paper 2017–04). INSEE.

Bernanke, Ben S. & Gertler, Mark & Gilchrist, Simon, 1999. The financial accelerator in a quantitative business cycle framework, in: J. B. Taylor & M. Woodford (ed.), *Handbook of Macroeconomics*, edition 1, volume 1, chapter 21, pages 1341–1393, Elsevier.

Blundell-Wignall, A., & Roulet, C. (2013). Bank lending puzzles: Business models and the responsiveness to policy. *OECD Journal: Financial Market Trends, 2013*(1), 7–30. https://doi.org/10.1787/fmt-2013-5k40m1nz55wj

Callonec, G., Landa, G., Malliet, P., Reynes, F., & Yeddir-Tamsamani, Y. (2013). *A full description of the Three-ME model: Multi-sector macroeconomic model for the evaluation of environmental and energy policy*. [Document de travail de l'OFCE]. OFCE.

de France, Banque, "Household Savings," Banque de France, 2021. https://www.banque-france.fr/fr/statistiques/epargne/epargne-des-menages-2021t4#:~:text=Sur%20l'ensemble%20de%20l,108%2C8%20en%202019

Drèze, J., & Modigliani, F. (1970). Consumption decisions under uncertainty. *Journal of Economic Theory, 5*(3), 308–335.

Ehrmann, M., Gambacorta, L., Martínez-Pagés, J., Sebestre, P., & Worms, A. (2001). *Financial systems and the role of banks in monetary policy transmission in the Euro area[2]* [Working Paper Series European Central Bank]. European Central Bank.

Gambacorta, L., & Marques-Ibanez, D. (2011). The bank lending channel: Lessons from the crisis: The bank lending channel: Lessons from the crisis. *Economic Policy, 26*(66), 135–182. https://doi.org/10.1111/j.1468-0327.2011.00261.x

Gollier, C. (2002). What does classical theory have to say about household portfolios? In L. Guiso, M. Haliassos, & T. Jappelli (Eds.), *Household portfolios* (pp. 27–54). The MIT Press. https://doi.org/10.7551/mitpress/3568.003.0005

Hörber, T. C., & Weber, G. (Eds.). (2022). *The European environmental conscience in EU politics: A developing ideology*. Routledge.

Ianc, N.-B., & Turcu, C. (2020). So alike, yet so different: Comparing fiscal multipliers across EU members and candidates. *Economic Modelling, 93*, 278–298. https://doi.org/10.1016/j.econmod.2020.07.018

Roger, S., & Vlček, J. (2011). *Macroeconomic costs of higher bank capital and liquidity requirements* (IMF Working Paper 11/103). IMF. https://papers.ssrn.com/sol3/papers.cfm?abstract_id=1847336

Stein, J. (1995). *An adverse selection model of bank asset and liability management with implications for the transmission of monetary policy* (w5217; p. w5217). National Bureau of Economic Research. https://doi.org/10.3386/w5217

7 Going Green? The Changing Role of the ECB in the Sustainability Strategy of the EU

Christoph Weber

Introduction

The European Central Bank (ECB) and its monetary policy have attracted attention from academics and NGOs for its role in the fight against climate change. Here are three prominent examples. Firstly, a public letter was sent to Christine Lagarde when she took office. The letter called on the ECB to "act now on climate change" and was supported by NGOs such as Greenpeace, Attac, the WWF European Policy Office, The Club of Rome, research institutes such as the Potsdam Institute for Climate Impact Research, and numerous individual scientists (Various, personal communication, 2019). Secondly, Greenpeace carried out two actions related to the ECB in 2021. On 10 March 2021, the ECB's headquarters in Frankfurt were conquered by Greenpeace. Greenpeace hung a banner on the ECB's building with the slogan "Stop financing climate killers!" and circulated with a flying banner with the slogan "act on climate now". This was accompanied by a report from the New Economics Foundation. The main criticism of this report is that the ECB's collateral framework is not in line with the goals of the Paris Agreement, and that it benefits carbon-intensive industries (Dafermos et al., 2021). On 23rd September 2021, Greenpeace placed ice blocks in front of the ECB's towers. The ice blocks showed the word future in German ("Zukunft"). Greenpeace also showed a banner urging the "ECB: Back our futures not polluters". Furthermore, the "ECB Listens" survey highlighted the public demand for other goals besides price stability. Civil society organisations such as Greenpeace encouraged their supporters to respond to the survey with standardised answers. When taking those answers into account, almost 50% mentioned climate change as an important secondary objective. Even when excluding non-original answers, the objective "climate change" still had strong support together with "growth and employment" (European Central Bank, 2021b).

This shows that the ECB is facing new demands they did not envision before. However, it shows that all EU institutions are now expected to contribute to the objective of climate neutrality. This could be an indication that a European environmental conscience has developed (Hörber & Weber, 2022). There are four selection criteria to determine if such a European environmental

DOI: 10.4324/9781032656359-10

conscience exists. Adapted to the case of the ECB and monetary policy, this chapter examines the following four selection criteria:

1 Is there a connection between monetary and environmental policies?
2 Is there an innovative nature of solutions, or a commitment to progress by the ECB?
3 Does the ECB have some expertise and therefore legitimacy for political leadership?
4 Is there some grassroots support and demand for effective solutions?

We will cover these four dimensions in this chapter. Firstly, we will discuss the link between monetary and environmental policies in the context of the transition towards a climate-neutral economy. The financial sector is crucial for funding this transition. At the same time, environmental regulations present some transition risks for enterprises and financial institutions. Secondly, the chapter will analyse the ECB's commitment and approaches of integrating climate change in its policies. Thirdly, it will compare the ECB with other central banks to determine leadership in the field of green central banking. Furthermore, the chapter will examine the compatibility of climate-related goals with the ECB's mandate. Fourthly, the chapter aims to find further evidence for public support and demand for an involvement of the ECB in the EU's green strategy.

There is a reason why the role of the ECB is so important in achieving environmental goals. The transition of the European economy to a zero-emissions economy requires major investments. As explained in the introduction to this book, these investments will have to come from the private sector, with only a small contribution from the public sector through public investments and subsidies. This private investment will have to be financed from other sources. For governments and larger companies, this can take the form of direct financing through green bonds, as explained in the chapter by Dejan Glavas and Florent Junior Okala Onana. However, households and micro-, small-, and medium-sized enterprises will have to rely mainly on indirect financing through commercial banks if they do not have sufficient savings. This brings the role of commercial banks to the fore. And they themselves are partly dependent on central banks and their liquidity. This is where the ECB comes in. Since the ECB provides liquidity to commercial banks, which then finance other projects, the ECB has an important indirect role in financing green projects. The ECB also acts as a buyer of government, corporate and green bonds. And the ECB requires collateral from commercial banks for providing them with liquidity, which in turn has an impact on asset markets. This shows that the ECB plays an essential role in the transformation of the EU economy, even if it is not obvious at first sight. The details of the ECB's involvement and importance are discussed in detail in this chapter.

The remainder of the chapter is structured as follows. Section 2 analyses the importance of climate stress tests. Section 3 deals with the issue of market neutrality in the case of monetary policy. Section 4 focuses on the large-scale asset-purchasing programmes implemented by central banks in the aftermath of the Great Recession. Section 5 compares central banks in terms of their engagement towards climate neutrality. Section 6 discusses whether climate-related goals are compatible with the ECB's mandate and if the ECB's mandate should be enlarged. Section 7 concludes.

Stress Tests: Climate Change and Climate Risk

Climate change presents considerable risks to life on Earth. However, its effects go beyond that and will impact various aspects, including the economy and the financial system. Battiston et al. (2017) demonstrated already in 2017 how direct and indirect exposure to climate change affects the financial system of the Euro system. For instance, they showed that European banks may face substantial losses in equity. This raises the question of how climate change and climate risk influence the financial sector.

First and foremost, any type of financial risk is relevant for central banks and monetary policy. As climate change poses substantial risks to financial stability, many scholars (e.g. Brunnermeier and Landau, 2020) express the need for central banks to integrate climate change into their financial risk assessment of monetary financial institutions. The issue of climate stress tests has attracted the attention of academic scholars leading to a debate about the implementation of such stress tests (Acharya et al., 2023). In fact, climate change can have sizeable effects on the financial system. One of many examples is the relation between the loss of biodiversity and financial stability (Salin et al., 2021). Carbon taxes represent another example. Empirical studies show how strong the effect of an introduction of a CO_2 tax at a level of 100 € would be on equity capital of commercial banks (Reinders et al., 2023).

When analysing the risks of climate change, we can differentiate between physical risk and transition risk. Physical risk includes the financial impact of changing climate (e.g., more frequent extreme weather events, rise of sea level, extreme heat, deforestation). Drastic increases in temperature rise could considerably reduce economic output. Several studies have estimated the impact of rising temperature on economic output. The estimates vary from 2.1% (Nordhaus, 2018), to 7–8% (Howard & Sterner, 2017) up to 23% (Burke et al., 2015) in the instance of global warming at a range of 3 degrees Celsius. In the end, these economic losses pose threats to financial stability.

Transition risks are the related to financial losses caused by adjustments towards lower-carbon economy and higher environmental sustainability (regulation or policy, technical progress). For example, certain natural resources must remain unused to reach climate goals. According to McGlade and Ekins (2015), 21% of the oil reserves, 6% of the gas reserves, and 89% of the coal reserves of Europe must remain unexploited to limit global heating to 2 degrees. The consequence is that companies owning those stranded assets suffer capital losses. Eventually, these losses may endanger financial stability (Campiglio et al., 2018).[1]

Moreover, there is an ongoing debate on whether climate catastrophes could trigger systemic financial crises. In analogy to Taleb's (2007) black swan (low probability events with extreme impact), this is called the green swan (Bolton et al., 2020). The issue is that climate change is related to deep uncertainty and high complexity with non-linear effects and tipping points. This makes the prediction of risk more challenging for financial institutions.

This explains the necessity of carrying out climate stress tests for commercial banks. Physical and transition risks affect the financial soundness of banks through their impact on credit risk, market risk, and liquidity risk (Acharya et al., 2023). Thus, climate stress tests examine how the economy and financial institutions are affected by climate change.

The ECB addressed this challenge by conducting its first climate risk stress test in 2021, which is based on the Network for Greening the Financial System (NGFS) framework. Firstly, the ECB conducted an economy-wide climate stress test in 2021. The stress evaluated the impact of climate change on the macroeconomy over the upcoming 30 years. One of the main findings was that it is better to act early as transition risks are lower than physical risks. Physical risks mainly affect the economy in the long run. The stress test emphasises that risk affection is different depending on the geographical regions as some areas of the Euro area are more exposed to climate change than others. Finally, the report concludes that climate change poses substantial systemic risks to banks and non-financial companies (Dunz et al., 2021).

Secondly, the ECB conducted another climate risk stress test in 2022 which confirmed the main findings of the stress test from 2021 but also revealing some new aspects. The stress test's main discovery was that commercial banks are still heavily relying on financing of carbon-intensive companies. At the moment, over 60% of the corporate interest rate payments that commercial banks receive from non-financial corporations come from the 22 highest air-polluting industries in Europe. Furthermore, roughly 60% of commercial banks have not yet implemented sound climate stress tests. Hence, the ECB wants to provide more guidance for commercial banks and best practices regarding different aspects of climate change adaptation (European Central Bank, 2022a).

Thirdly, the ECB conducted a climate risk stress test on the balance sheet of the Eurosystem in 2022. The results display the impact of both physical and transition risk on the Eurosystem balance sheet. In general, physical risks are more significant than transition risks. Furthermore, the report shows that an orderly transition of the economy poses the lowest risk among the three scenarios (orderly transition, disorderly transition, hot house scenario). Compared to an orderly transition, the risk is 20% higher for a disorderly transition and 30% higher for the hot house scenario (i.e., no adoption of any climate policies) (Germann et al., 2023).

This highlights the link between the financial system and environmental policies. The aim of environmental policies is to achieve an orderly transition of the economy, which necessitates substantial financial investments across various sectors of the economy. On the other hand, environmental regulations also affect financial institutions and investors. Nonetheless, climate change poses significant financial risks. As shown by empirical studies, it is preferable to promote a transition of the economy. While this does carry some financial risk, it is still lower than taking no action. Additionally, the ECB is developing some expertise in conducting climate stress tests. These are both indications of a growing European environmental conscience.

Markets Neutrality

Another important aspect of green monetary policy is market neutrality. Market neutrality would require the ECB to hold assets in proportion to their market shares. The ECB's main collateral requirements are that the assets have an investment grade rating, a maturity of 1-31 years, and are issued by non-bank entities established in the European Economic Area. These requirements can be challenging for green bonds, especially the investment grade rating. As a result, green bonds may be at a disadvantage compared to brown bonds when it comes to serving as collateral. This explains the criticism of the ECB that its collateral framework favours brown bonds.

One of the decisions taken by the ECB in this regard was to accept sustainability-linked bonds from the beginning of 2021 onwards. In their chapter in this book, Dejan Glavas and Florent Junior Okala Onana explain in detail how the green bond market has developed and how it works. This applies to bonds linked to either the EU Taxonomy Regulation or (one of) the Sustainable Development Goals (European Central Bank, 2020). The provisions of the EU taxonomy are discussed in detail in Chiara Pappalardo's chapter in this book. Another future step will be to make "climate-related corporate disclosures compulsory for bonds to remain eligible as collateral" (Schnabel, 2023). Finally, the ECB will adjust its collateral framework by limiting "the share of assets issued by entities with a high carbon footprint that can be pledged as collateral by individual counterparties when borrowing from the Eurosystem" (European Central Bank, 2022b). The ECB's adjustments are showing early signs of success, with the proportion of green bonds in the ECB's portfolio increasing from 13% to 20% between 2022 and 2023, with a target of 25% in 2024 (European Central Bank, 2024c).

Schnabel (2021b) argues that the ECB should aim for market efficiency and not just market neutrality. Market efficiency would require monetary policy to take into account the costs and risks associated with climate change. For example, in June 2018, the People's Bank of China gave green bonds a "first-among-others" status. This had a positive impact on green bonds as they experienced a 46 basis point increase in value thanks to the "first-among-equals" status (Macaire & Naef, 2023).

Another related issue is the size of the haircut of green bonds relative to brown bonds. Giovanardi et al. (2023) analyse the different treatment of green and brown bonds in a DSGE model. They find an optimal haircut difference of 20 percentage points between green bonds and conventional bonds. However, the overall impact on the macroeconomy would be very limited. In the extreme case of no haircut for green bonds, green corporate investments would only increase by only 0.6% compared to the case of equal treatment of green and brown bonds. Giovanardi et al. (2023) therefore conclude that a preferential treatment for green bonds should only be implemented if countries do not introduce significant carbon taxes.

Again, we find some evidence of a developing European environmental conscience, as there is a clear link between monetary and environmental policy. The ECB makes use of the EU Taxonomy regulation, an area where the EU has developed some expertise in setting standards for markets. So, the ECB has responded to the criticism and is reviewing its collateral framework.

Quantitative Easing and Purchase of Corporate Bonds

Apart from the fact that the collateral framework favoured brown industries, the ECB was also criticised for its purchases of corporate bonds. The first Covered Bonds Purchase Programme (CBPP1) was launched in 2009 and had a volume of 60 billion euros. Remaining holdings amount to less than 400 million euros. The second Covered Bonds Purchase Programme (CBPP2) was launched in November 2011 with a volume of 16.4 billion euros (the initial target was 40 billion euros). In March 2015, the ECB launched an Expanded Asset Purchase Programme (APP). The APP included purchases of asset-backed securities (ABSPP), covered bonds (CBPP3), public sector bonds (PSPP), and corporate bonds (CBPP). Monthly net purchases varied between 15 and 80 billion euros. The total volume of assets in the APP reached a maximum of 3,265 billion euros[2]. In addition, the ECB launched a Pandemic Emergency Purchase Programme (PEPP) in March 2020. This added a further 1,850 billion euro of asset purchases.

The main criticism relates to corporate bonds. The maximum amount of CBPP holdings in the APP was 345 billion euro. In addition, the ECB has corporate bond holdings of 46 billion euros from the PEPP. As Papoutsi et al. (2021) explain there is a bias in ECB holdings towards brown industries. For instance, there were special regulations for eligibility in the CSPP of the ECB. The regulations include denomination in euros, issuance by a non-financial company from the euro area, investment grade rating, and eligibility as collateral for Eurosystem operations (Matikainen et al., 2017). For the year 2017, Matikainen et al. (2017) estimate that the eligibility criteria reduce the pool of corporate bonds from potentially 80,000 to roughly 1,150. In the end, the ECB bought corporate bonds from around 850 companies. The ECB bought predominantly bonds from the manufacturing and utilities sector which are responsible for roughly 60% of the greenhouse emissions but only 18% of GDP. In contrast, sectors such as real estate, wholesale, and retail were underrepresented compared to their contribution to GDP. Additionally, the ECB purchased corporate bonds solely from gas and oil-producing companies when only considering energy sector bond purchases (Matikainen et al., 2017). Therefore, the main criticism is that the ECB buys corporate bonds from highly polluting companies. To give a few examples, the ECB holds corporate bonds from ArcelorMittal Belgium (iron and steel production), Ryanair, Deutsche Lufthansa AG (airline industry), ENEL, Engie, OMV, RWE, Shell, Total (energy sector) and BMW, Daimler, Stellantis, Volkswagen (car manufacturers).

This leads to the question of how to select the assets for the portfolio. Amel-Zadeh and Serafeim (2018) distinguish between eight main strategies[3]. For central banks, the most promising options are either negative screening (not buying assets from "bad" companies) or portfolio tilting (buying disproportionately bonds that perform well on one of the ESG aspects). Schoenmaker (2021) has proposed a portfolio-tilting approach for the ECB. In a numerical example, he shows how the ECB could reduce the carbon intensity of its portfolio by 41%, 55%, or 69% by applying low, medium, or high tilts respectively.

The ECB has responded to the criticism and adjusted its approach accordingly. The first decision was to take climate risks into account when reinvesting in corporate bonds from October 2022 onwards.

Since February 2023 corporate bond purchases have been tilted towards companies with a better climate performance. The ECB also calculates the carbon footprint of its corporate bond holdings. Table 7.1 shows how the ECB's carbon footprint and carbon intensity have evolved over time (European Central Bank, 2024b). The ECB now calculates the footprint of 99% of its asset holdings, with the exception of asset-backed securities (European Central Bank, 2024c). It shows that the size of the portfolio under the APP and PEPP increased between 2018 and 2022, while it decreased in 2023 following the ECB's decision to begin quantitative tightening. The associated total carbon emissions increased between 2018 and 2022 but started to decrease in 2022. At the same time, the carbon footprint and carbon intensity have decreased each year. Overall, the carbon footprint has decreased by 45% and the carbon intensity by around 50%. Of course, part of this development can be explained by

Table 7.1 Carbon footprint of corporate bonds holdings

	2018	*2019*	*2020*	*2021*	*2022*	*2023*
Portfolio size (bn.€)	173	180	288	346	385	367
Total carbon emissions	38	32	47	55	50	47
(Scope 1 & 2 emissions in mega tons CO_2 emitted)						
Carbon footprint	240	195	172	166	135	132
(Tons of CO_2 emissions per mio. € invested)						
Carbon intensity	372	332	311	295	196	187
(Tons of CO_2 emissions per mio. € revenue)						

Data Source: European Central Bank (2024b)

the fact that the EU's total greenhouse gas (GHG) emissions went down by 14% between 2018 and 2023. However, the reduction in the carbon footprint is greater than the overall reduction in GHG emissions. This does not mean that the ECB has already reached its target as it still holds bonds from "brown" companies, as explained above. However, Schnabel (2023) argues that the ECB "should not divest completely, at least not initially, from those companies whose actions are particularly important in managing the green transition, but rather foster incentives for them to reduce emissions further" (Schnabel, 2023). The current approach is considered insufficient by Greenpeace as the tilt is only employed temporarily (for 5 months) and the CSPP is then abandoned (Dafermos et al., 2023). However, the ECB has announced that it plans to reduce the carbon footprint of its corporate bond holdings by 7% per year in the coming period (European Central Bank, 2024c).

Despite the criticism from Greenpeace, the ECB's approach is a promising way forward, as the inclusion of green bonds has a positive impact on financing conditions for green projects. Bremus et al. (2021) showed that eligible green bonds benefited from a reduction of 18 to 33 basis points compared to non-eligible green bonds in the case of the CSPP and the PEPP.

With regard to the PSPP, which consists mainly of government bond holdings, the ECB has not yet taken any measures to take carbon footprints into account. Schnabel (2023) argues that there are several problems with setting ESG rules for public sector bond purchases. The first problem is that the government bond purchases are usually based on the ECB's capital key. Secondly, it is difficult to determine how to assess whether and which government bonds contribute to the goals of the Paris Agreement. Thirdly, the green bond market is very limited compared to the sovereign bond market.

Again, we have found some evidence of a link between monetary and environmental policy. The ECB is committed to progress and is building expertise in this area. This is what NGOs such as Greenpeace are demanding. Thus, all four dimensions of the European Environmental Conscience can be validated here.

Comparison With Other Central Banks

To better assess the ECB's position among its peers and to determine whether the ECB is a leader in green central banking, we compare the ECB with other selected central banks. Our approach is mainly to assess the involvement in the NGFS, a leading consortium for collaboration on greening monetary policy. In contrast, the role of the International Monetary Fund (IMF) has so far been very limited, although the IMF has started to work on climate change, albeit in a very limited way (Committeri et al., 2022). We therefore focus mainly on the NGFS.

The NGFS is a pool of 125 members trying to green the financial system. It was founded in December 2017 by eight central banks including the central banks of China, England, France, Germany, Mexico, the Netherlands, and

Singapore as well as the Swedish financial supervision. The number of countries not represented in the NGFS is very limited and includes the Czech Republic, Ecuador, Gabon, Guatemala, Guinea, Guinea-Bissau, Honduras, Jamaica, Nicaragua, Qatar, Senegal, and Venezuela.

There are a wide variety of different positions on the role of climate change for central banks. Based on this observation, my approach is to classify central banks into three types of categories according to their position on climate change: non-active, reactive, and proactive. It is interesting to note that the mandates of central banks do not necessarily correspond to their category. Dikau and Volz (2021) present a survey of central banks. They analyse 125 central banks. Of these, 70 central banks have mandates that directly or indirectly mention sustainability. 15 central banks have an explicit goal related to sustainability. One of them is the Czech Republic's central bank. However, the Czech central bank is not a member of the NGFS. On the other hand, there are 33 central banks with a mandate that does not mention sustainability. However, some of them are very active in the fight against climate change, such as China, India, or Japan. We therefore differentiate central banks on the basis of their actions, rather than their mandates.

The first group consists of central banks that do not believe that climate change should be part of central banking. For instance, Jiří Rusnok, Governor of the Czech national bank, described the Czech central bank's position on climate change by pointing to "many big risks associated with this approach." According to him,

> environmental pollution is a typical example of market failure, and central banks should not be at the forefront of efforts to correct such failures. It's not their job. What's more, going green implies abandoning market neutrality, a principle which ensures maximum monetary policy effectiveness

(Rusnok, 2021). This argument is questionable as many scholars point to the non-market neutrality of traditional monetary policy. However, the position apparently seems to see market neutrality in a sense where market interventions are not justified and market outcomes should be accepted by the central bank. This can also be criticised, as central banks are always market makers, setting, for example, interest rates and thus influencing the market equilibrium. Overall, the central banks in this group are very heterogeneous. Their position of non-action refers to the fact that they do not see monetary policy as a tool for going green. However, there are still some initiatives towards environmental sustainability such as planting of trees by the Central Bank of Honduras (2021).

The second group of countries includes central banks with a reactive stance. In this perspective, central banks are not actors in the fight against climate change, but rather react to climate change as it affects the financial system. However, they may still see a role for central banks. For example, the Bank of Canada believes that central banks "in a unique position to improve society's understanding of the economic and financial system impacts of climate change and the policies to address it" (Bank of Canada, n.d.). Furthermore, the Bank of Canada aims to "better understand climate change because of its important effects on the economy and prices" (Bank of Canada, n.d.).

The third group concerns central banks that can be considered to be proactive. The Swedish central bank is at the forefront of these activities. This starts with the fact that the Swedish central bank has been calculating the carbon footprint of its foreign exchange reserves since 2017 (Brattström & Gajic, 2022). One of the results of assessing the carbon footprint of reserve holdings was that the Swedish central bank sold bonds of Alberta (Canada), Queensland, and Western Australia (both Australia). Similarly, the Swedish Riksbank measures the carbon footprint of its corporate bond holdings and only buys bonds from companies that respect "international standards and norms for sustainability" (Sveriges Riksbank, 2020). This ultimately means that central banks are perceived as being able "to correct the carbon bias in capital markets" (Knot, 2021). In the future, sustainability could be a fourth aspect to be regularly considered for foreign exchange reserve holdings, in addition to liquidity, safety, and return (Fender et al., 2020)[4]. Another aspect is the implementation of the Paris Agreement Capital Transition Assessment (PACTA). For instance, PACTA has been used by the Bank of England, the Swedish Riksbank, the Dutch central bank and others.

Some central banks in this group are trying to stimulate green investments directly. One such approach is the introduction of dual green interest rates. The idea would be to have one interest rate for conventional (or brown) investments and one interest rate for green investments. Since central banks do not usually lend money directly to non-financial corporations, this would rather mean that the central bank offers liquidity to commercial banks at two different interest rates, depending on whether they use this liquidity to finance brown or green projects. Providing liquidity at lower interest rates to commercial banks that finance green projects would be one way to stimulate the environmental transition. For example, the Bank of Japan offers "Funds-Supplying Operations to Support Financing for Climate Change Response". These are special loans offered every 6 months at an interest rate of 0%. Some central banks, such as the Reserve Bank of India and the Bangladesh Bank, require commercial banks to lend a minimum amount of funds to companies in the environmental sector. Other central banks, such as the Banque du Liban, require commercial banks to hold lower minimum reserves if they favour loans for renewable energy and energy efficiency (Campiglio et al., 2018). However, despite political support from French President Macron, the ECB has so far shied away from green dual rates. The ECB's refusal to move in this direction has attracted some criticism (Woolley, 2024).

However, the fact that central banks in this category are proactive when it comes to environmental sustainability does not mean that they see monetary policy or central banks at the forefront of the fight against climate change. As the Swedish central bank puts it, a central bank "can contribute the means available within its mandate, and as a complement to other policy. However, the most effective measures to limit climate change fall within the remit of other policy areas" (Sveriges Riksbank, 2022). The ECB also shares this view, stating that "governments and parliaments have the primary responsibility to act on climate change" (European Central Bank, 2021c).

To be more precise about the ECB's approach, let us look at the "Detailed roadmap of climate change-related actions" published by the European Central Bank (2021a). This action plan and roadmap were a part of the ECB's strategy review. The roadmap outlines the various steps that will be taken to integrate environmental considerations into the ECB's monetary policy:

- Adjustment of macroeconomic models and additional research (2022–2024)
- Strengthened assessment of risk of assets related to the climate as well as the carbon footprint of financial corporations (2021–2022)
- Disclosure requirements for assets (2021–2024)
- Climate stress test (2022)
- Additional criteria for corporate bonds purchases in line with Paris Accord (2021–2024)
- Market neutrality of bond purchases (2021–2022)

The next phases have been outlined as a follow-up to the 2022 Climate Agenda. The focus in 2024 and 2025 will be on "the implications of the green transition, the physical impact of climate change, and the risk that nature loss and degradation pose to the economy" (European Central Bank, 2024a). In addition, the ECB established a Climate Change Centre (European Central Bank, 2021c). The ECB also became an observer of the EU's Platform on Sustainable Finance which was initially launched in 2020 for a two-year period but has been renewed with a new composition for 2023.

Compared to other central banks in the world, the ECB ranks as one of the more proactive central banks. The ECB has gained some expertise and therefore legitimacy to take a leading role in green central banking. It is true that the ECB does not go as far as some other central banks, which have introduced additional instruments such as green loans (Bank of Japan), minimum lending requirements for environmental projects (Reserve Bank of India), or lower minimum reserve requirements for banks lending to projects related to renewable energy and energy efficiency (Banque du Liban). However, this can be explained by the fact that Europe can rely on the European Investment Bank (EIB). In conclusion, the ECB is active within the limits of its mandate. We can see that the ECB is committed to progress in line with a developing European Environmental Conscience.

The ECB's Mandate

Finally, we want to assess whether and to what extent tackling climate change is in line with the ECB's mandate. It is important to remember that central banks have "unelected power", as Tucker (2018) explains in detail in his book.

This stems from the fact that most central banks are granted independence from political influence. The result is that, even in a democratic system, there is only indirect and very marginal democratic legitimacy for central banks. The only democratic elements are that the objectives of monetary policy are set by parliaments in laws or, in the case of the ECB, in the TFEU. The other democratic aspect is that central bankers are elected or appointed by elected politicians. There is also a degree of accountability, as central banks have to report to parliaments. However, the direct influence of the public on monetary policy decision making is almost nil, as citizens cannot vote on the heads of central banks and changing monetary policy goals is usually a lengthy process. At the same time, central banks usually have only a small number of objectives (including price stability, economic stability, low unemployment, financial stability). In the case of the ECB, there is only one primary objective according to Article 127(1) TFEU, namely price stability. However, the ESCB also has to support the general economic objectives of the EU, as set out in Art. 3 TEU. According to Art. 127(6) TFEU, the ECB may be assigned additional tasks. This happened in the past when the ECB was asked to take over banking supervision as part of the Single Supervisory Mechanism in 2013. Another possibility for adaptation is provided by Art. 129(3) TFEU, which allows the European Parliament and the Council to amend the statute of the ECB / ESCB.

This raises three main questions:

1 Are climate change related objectives compatible with the current mandate of the ECB?
2 Should the ECB's mandate be extended?
3 Does the inclusion of climate change objectives jeopardise the independence of the ECB?

The first main question concerns the compatibility of climate change-related goals with the ECB's mandate. ECB Executive Board member Elderson (2021) explained in a blog, which was also published in various newspapers, why he believes that tackling climate change is already part of the ECB's mandate. The first aspect he mentions is that climate change can have an impact on price stability, which is the primary objective of the ECB. In addition, structural change can affect financial stability. Furthermore, he argues that the "Court of Justice of the European Union has confirmed that catering for the preconditions required for the pursuit of our primary objective falls within our mandate to maintain price stability" (Elderson, 2021).

Indeed, climate change may affect macroeconomic variables (inflation, output) and the ECB's primary objective (price stability). And according to Article 128(3) TFEU, the main principles are stable prices, sound public finances and monetary and sustainable balances of payments. One concern could be that there is a trade-off between climate policy and monetary policy, as CO_2 prices affect consumer price inflation. There is some empirical evidence that higher EU allowance prices contribute to higher inflation expectations and inflation in the euro area (Nishigaki, 2023). Higher CO_2 prices also lead to higher inflation expectations among firms, as shown by Hensel et al. (2024) for the case of France. However, recent empirical studies show that an increase in the CO_2 price per tonne from €85 in 2021 to €140 in 2030 would have an impact of no more than 0.2 percentage points on annual inflation rates (Brand et al., 2023). Konradt et al. (2024) come to similar or slightly higher estimates. They predict that an increase in the CO_2 price to 150 euros per tonne would add 0.2 to 0.4 percentage points to the inflation rate. But the truth is that CO_2 prices should be much higher if they were to reflect the social costs. For example, Rennert et al. (2022) argue that the social cost of CO_2 is around USD 185 per tonne.

Another facet of the discussion is that the equilibrium real interest rate may be affected. It is necessary to assess whether climate change leads to a change in potential economic growth (Brunnermeier & Landau, 2020). For example, climate change could affect labour productivity or destroy factors of production, resulting in lower economic growth. The result could be a lower equilibrium real interest rate (Schnabel, 2021a). This would directly affect the role of monetary policy in setting nominal short-term interest rates.

The second aspect explained by Elderson (2021) is that the ECB is required to support general economic policy. In fact, the ESCB also has to support the general economic objectives of the EU, which is laid down in Art. 3 TEU. And Art. 3.3 TEU explicitly mentions as an objective of the EU "a high level of protection and improvement of the quality of the environment". This explains the reaction of the European Parliament. In a resolution of 12 February 2020, the European Parliament "recalls that, as an EU institution, the ECB is bound by the Paris Agreement on climate change and that this should be reflected in its policies, while fully respecting its mandate and its independence" (European Parliament, 2020). They also called on the ECB to

> implement the environmental, social and governance principles (ESG principles) into its policies, while fully respecting its mandate and its independence (European Parliament, 2020). In addition, the Committee on Economic and Monetary Affairs (2020) called on the ECB "to align its collateral framework with climate change-related risks and to disclose its level of alignment with the Paris Agreement.

All in all, the main aspects mentioned above are uncontroversial. To sum up: as long as the ECB maintains its primary focus on price stability, climate change is already covered by the ECB's mandate. And even the ECB has a limited impact on the environmental transition. For example, the announcement of the ECB's review of its monetary policy strategy in 2021 has already had two positive effects on the green bond market. Firstly, it led to a reduction in the yields to maturity of green bonds. Secondly, it encouraged an increase in green bond issuance (especially investment-grade green bonds) (Eliet-Doillet & Maino, 2022).

The second main issue concerns a change in the ECB's mandate. Changing the objectives or emphasising certain aspects would not be entirely unprecedented if we look back in history. As Goodhart (2011) shows, there have been various episodes of monetary policy objectives, although the primary focus has been on price and financial stability. However, as we have already discussed, climate change may have some implications for financial stability.

It is undeniable that monetary policy, and in particular quantitative easing, has an impact on green bond markets. Hilmi et al. (2022) show that quantitative easing (proxied with M3) significantly increases the volume of green bonds and the likelihood of green bond issuance. Another question is whether the ECB should go further and engage in green quantitative easing. Some scholars, such as Grauwe (2019), recommend this. However, the question is whether it really makes sense to conduct such green QE, given the small impact it would have. Abiry et al. (2022) estimate the effects of green QE. They conclude that green QE would only reduce temperature increase by 0.04°C by 2100. One of the reasons for this small effect is that green QE would crowd out private investment in green projects. A carbon tax is much more effective in achieving environmental goals. Ferrari and Landi (2023) come up with a similar conclusion: green QE has a positive, but very small impact on pollution. In the long-run, green QE has no effect. And commercial banks adjust their portfolios anyway thanks to important international agreements such as COP21 (Reghezza et al., 2022). Thus, the main strategy to increase the supply of credit for green projects and companies is not QE, but it is regulation that encourages green innovation and stops subsidising polluting industries.

It is therefore better not to add additional objectives for the ECB in the context of climate policy. Firstly, the ECB's impact on climate change is very limited. The main role in the fight against climate change is reserved for economic and fiscal policy. And according to Art. 120 TFEU, economic policy is a matter for the member states. Giving central banks explicit targets for climate change would add a burden and raise expectations that they cannot meet (Hansen, 2022). Central banks cannot achieve far-reaching climate goals because they lack the means and instruments to do so. Monetary policy instruments are designed for price stability, not for combating climate change.

It is also better to keep the ECB out of direct financing of green projects, similar to the approach taken by the Bank of Japan. A better solution is a division of labour between the EIB and the ECB. Direct lending to green projects would reduce the ECB's financial independence. The EIB is already a leader in financing climate projects. In 2022, 56% of the EIB's lending went to projects related to climate action and environmental sustainability. For the decade 2020 to 2030, the EIB aims to lend 1 trillion euros in support of climate action and environmental sustainability projects (EIB, 2023). Secondly, any change in the mandate would have to be democratically legitimised. As Volz (2017) elaborates, central bankers are not elected and therefore a debate in society about climate policy is inevitable. Thirdly, the ECB should not be overburdened with more and more goals. The ECB is the main actor in maintaining price stability. Moreover, the

ECB has provided an implicit guarantee for the euro. Many national central banks in the euro area are already suffering losses due to the large asset purchase programmes combined with higher interest rates. These losses are likely to continue for several years. Giving the ECB more targets would require it to take on more risks, with potential financial ramifications.

The third question is whether an inclusion of the ECB in the fight against climate change jeopardises its central bank independence. There are two opposing positions on this point. On the one hand, Cochrane (2020) argues that adding climate goals destroys independence. Hansen (2022) also raises concerns that an extension of the central banks' mandates could eventually endanger their independence. On the other hand, Honohan (2019) postulates that independence is endangered if central banks ignore their secondary targets. This is also the position of Fisher and Murphy (2021) who argue that central banks could lose their independence if they do not consider climate for monetary policy.

It is important to remember that it is not up to the ECB to decide which objectives it should pursue. The primary objectives are laid down in Art. 127 TFEU. Changing them requires a democratic decision. So, the real question is whether the European Council and the European Parliament should give the ECB an explicit objective to address climate change. As mentioned above, the ESCB has to "support the general economic policies in the Union with a view to contributing to the achievement of the objectives of the Union as laid down in Article 3 of the TEU" (Art. 127.1 TFEU). And "a high level of protection and improvement of the quality of the environment" is one of the objectives set out in Art. 3.3 TEU. Therefore, there is no need to adapt the objectives of the ECB. Quite the contrary: as noted above, forcing the ECB to be more active, for example, by asking it to conduct green QE would further threaten the financial independence that has been a concern since the Great Recession (Weber & Forschner, 2014). However, it is important to emphasise that the ECB and the NCBs may not seek instructions, nor may anyone give the ECB and the NCBs instructions according to Art. 130 TFEU. While exchanging with the public and the European Parliament is essential for the ECB, it is the ECB alone that decides how to conduct monetary policy in the euro area. As long as this rule is respected, the ECB's independence is not compromised.

Another independence-related concern was put forth by German economist Wieland who argued that it would be difficult to exit from quantitative easing if the ECB started buying green bonds (Siedenbiedel, 2021). This concern has already been allayed. In June 2022, the Governing Council of the ECB decided to stop net asset purchases under the APP from July 2022. From October 2022, the ECB started to take climate risks into account when reinvesting corporate bonds. Since February 2023, the ECB's corporate bonds purchases are tilted towards companies with a better environmental performance. And since March 2023, the ECB has been reducing its holdings in the APP by an average of 15 billion euros per month. So, although the ECB has introduced green rules for corporate bonds, it has not prevented the ECB from tapering.

Another concern is that there is some evidence that higher inflation over a longer period could lead to lower CO_2 emissions (Grolleau & Weber, 2024). If this were to lead the ECB to compromise its inflation target in the hope of helping the environment, this would be questionable and not supported by its mandate. However, there is no evidence that the ECB is moving in this direction.

The opposing view was that the ECB's independence could be at risk if it ignored climate change. This argument has some validity. Even if central banks are largely independent, they are held accountable for the results of their monetary policy. The ECB recognises this: it considers itself "accountable first and foremost to EU citizens, and—more formally—to the European Parliament, which is the only European institution directly elected by EU citizens (European Central Bank, 2011, p. 86). A monetary policy that runs counter to other EU initiatives would put the status of the ECB at risk. Trust and credibility are essential for the ECB as the ECB is not directly democratically legitimised (Weber, 2014).

We can conclude that tackling climate change within the limits of the ECB's mandate would not threaten its independence. On the contrary, failing to respond to public demands could jeopardise the ECB's status.

In a recent study, Aguila and Wullweber (2024) analysed how the ECB explains its focus on climate change. Members of the ECB's Executive Board typically refer to three main reasons to justify the ECB's focus on climate change. The first is that climate change poses financial risks with potential implications for banks and the financial system. The second is that large investments are needed to achieve climate neutrality - as discussed in the introduction and conclusions of this book. The third reason given by the Executive Board of the ECB is that climate change may affect price stability, which is at the core of the ECB's mandate (Aguila & Wullweber, 2024). This is in line with the ECB's mandate and responsibilities. Neglecting the role of a central bank would not be justified. This does not mean that the ECB has to play an active role in the fight against climate change, but it can do its fair share within its existing mandate without compromising its independence.

Conclusions

The main objective of this chapter was to analyse whether we can identify a European environmental conscience in the case of the ECB. The European environmental conscience was proposed by Hörber and Weber (2022) on the basis of four criteria. We have adapted the four criteria to the case of monetary policy.

The first aspect is the connection between monetary policy and environmental policies. We have seen that climate change poses a threat to financial systems. At the same time, the environmental transition requires financing. The financial sector will therefore play an important role in enabling the environmental transition. Furthermore, the issue of market neutrality underlines the link between monetary and environmental policy. Ignoring market neutrality undermines efforts to help green projects gain market access. Finally, environmental regulation affects monetary policy mainly through changes in the price of CO_2. To make the EU Emissions Trading System a success, the ECB will have to deal with price increases for some goods and services while maintaining overall price level stability. These considerations underline the importance of monetary policy and the ECB in the EU's green strategy.

The second dimension is the ECB's role as a leader. Section 5 has shown that the ECB is one of the most committed central banks when it comes to tackling climate change and integrating climate aspects into monetary policy and financial supervision. It is true that some other central banks go even further but as we discussed in the context of the ECB's mandate, the ECB must respect certain limits. Three euro area national central banks were among the founding members of the NGFS. This shows that they were ahead of the curve in recognising the importance of climate change for central banks and the financial system.

The third aspect is whether the ECB is building up expertise in this field and thus gaining legitimacy for political leadership. Globally, the answer to this question is also yes. For example, the ECB conducted its first climate risk stress test in 2021. The stress test was based on the NGFS framework, notably with some involvement of the ECB and some euro area central banks.

The final question about the European environmental conscience is whether there is grassroots support and demand for effective solutions. The ECB was urged to "act now on climate change" by NGOs such as Greenpeace, Attac, the Club of Rome and others in 2019. This was followed by public Greenpeace protests against the ECB's monetary policy in 2021. Finally, the "ECB Listens" survey showed public support for addressing climate change as part of the ECB's mandate. Thus, there is public support and demand for finding solutions for greening monetary policy and the financial system.

In essence, we found some evidence that a European environmental conscience is present also in the case of the ECB. Thus, the European environmental conscience seems to spread from traditional policy areas such as energy and security to other areas such as monetary policy and the financial sector. This is an important development as the environmental transition of the economy requires an involvement of all key actors.

Notes

1 As a solution, one report supported by NGOs such as Reclaim Finance or Rainforest Action Network calls for the ECB to serve as a bad bank taking over stranded assets to reduce the transition risks of commercial banks (Giraud et al., 2021).

2 End-of-month book value at amortised cost.

3 These strategies by professional investors include full integration into individual stock evaluation, engagement/active ownership, positive screening, risk factor/risk premium investing, relative screening/best-in-class screening, thematic investing, overlay/portfolio tilt, and negative screening (Amel-Zadeh & Serafeim, 2018).

4 These 3 principles are the only objectives for the ECB's foreign reserves (European Central Bank, 2023). To date, the ECB has not applied any other restrictions in terms of ESG criteria or ethical restrictions to its foreign exchange reserve management.

References

Abiry, R., Ferdinandusse, M., Ludwig, A., & Nerlich, C. (2022). Climate change mitigation: How effective is green quantitative easing? *ZEW-Centre for European Economic Research Discussion Paper*, *22–027*.

Acharya, V. V., Berner, R., Engle III, R. F., Jung, H., Stroebel, J., Zeng, X., & Zhao, Y. (2023). *Climate stress testing*. National Bureau of Economic Research.

Aguila, N., & Wullweber, J. (2024). Legitimising green monetary policies: Market liberalism, layered central banking, and the ECB's ongoing discursive shift from environmental risks to price stability. *Journal of European Public Policy*, 1–32. https://doi.org/10.1080/13501763.2024.2317969

Amel-Zadeh, A., & Serafeim, G. (2018). Why and how investors use ESG information: Evidence from a global survey. *Financial Analysts Journal*, *74*(3), 87–103.

Bank of Canada. (n.d.). *Climate change is a big issue for central banks*. Retrieved 15 May 2023, from https://www.bankofcanada.ca/2019/11/climate-change-is-a-big-issue-for-central-banks/

Battiston, S., Mandel, A., Monasterolo, I., Schütze, F., & Visentin, G. (2017). A climate stress-test of the financial system. *Nature Climate Change*, *7*(4), 283–288.

Bolton, P., Despres, M., Da Silva, L. A. P., Samama, F., & Svartzman, R. (2020). *The green swan. BIS Books*.

Brand, C., Coenen, G., Hutchinson, J., & Saint Guilhem, A. (2023). *How will higher carbon prices affect growth and inflation?* https://www.ecb.europa.eu/press/blog/date/2023/html/ecb.blog.230525~4a51965f26.en.html

Brattström, & Gajic, R. (2022). *The carbon footprint of the foreign exchange reserves has decreased over the last five years*. https://www.riksbank.se/en-gb/press-and-published/publications/economic-commentaries/the-carbon-footprint-of-the-assets-in-the-riksbanks-foreign-exchange-reserves/the-carbon-footprint-of-the-assets-in-the-riksbanks-foreign-exchange-reserves/the-carbon-footprint-of-the-foreign-exchange-reserves-has-decreased-over-the-last-five-years/

Bremus, F., Schütze, F., & Zaklan, A. (2021). *The impact of ECB corporate sector purchases on European green bonds*. DIW Berlin, German Institute for Economic Research.

Brunnermeier, M., & Landau, J.-P. (2020). Central banks and climate change. *Disponible En VoxEU. Org*: https://Voxeu.Org/Article/Central-Banks-and-Climate-Change

Burke, M., Hsiang, S. M., & Miguel, E. (2015). Global non-linear effect of temperature on economic production. *Nature*, *527*(7577), 235–239.

Campiglio, E., Dafermos, Y., Monnin, P., Ryan-Collins, J., Schotten, G., & Tanaka, M. (2018). Climate change challenges for central banks and financial regulators. *Nature Climate Change*, *8*(6), 462–468.

Central Bank of Honduras. (2021). *Boletín De Prensa No. 34/2021*. https://www.bch.hn/administrativas/RI/Enlaces%20Descarga%20Boletines%20de%20Proyeccin%20Institucio/Bolet%C3%ADn%20de%20Prensa%20No%2034-2021%20Banco%20Central%20de%20Honduras%20conmemora%20el%20d%C3%ADa%20del%20%C3%A1rbol.pdf#search=cambio%20climatico

Cochrane, J. H. (2020). *Central banks and climate: A case of mission creep*. Hoover Institution. https://www.hoover.org/research/central-banks-and-climate-case-mission-creep

Committee on Economic and Monetary Affairs. (2020). *Report on the European central bank annual report 2020*. https://www.europarl.europa.eu/doceo/document/A-9-2021-0002_EN.html

Committeri, M., Brüggemann, A., Kosterink, P., Reininger, T., Stevens, L., Vonessen, B., Zaghini, A., Garrido, I., Meensel, L. V., & Strašuna, L. (2022). The role of the IMF in addressing climate change risks. *ECB Occasional Paper*, *2022/309*.

Dafermos, Y., Gabor, D., Nikolaidi, M., Gogolewski, J., & Vargas, M. (2023). *Broken promises: The ECB's widening Paris gap.*

Dafermos, Y., Gabor, D., Nikolaidi, M., Pawloff, A., & van Lerven, F. (2021). *Greening the Eurosystem collateral framework: How to decarbonise the ECB's monetary policy.*

Dikau, S., & Volz, U. (2021). Central bank mandates, sustainability objectives and the promotion of green finance. *Ecological Economics, 184*, 107022.

Dunz, N., Emambakhsh, T., Hennig, T., Kaijser, M., Kouratzoglou, C., & Salleo, C. (2021). ECB's economy-wide climate stress test. *ECB Occasional Paper, 2021/281.*

EIB. (2023). *Climate action and environmental sustainability overview 2023.* https://www. eib.org/attachments/lucalli/20220287_climate_action_and_enviromental_ sustainability_overview_2023_en.pdf

Elderson, F. (2021). *Greening monetary policy.* https://www.ecb.europa.eu/press/blog/ date/2021/html/ecb.blog210213~7e26af8606.en.html

Eliet-Doillet, A., & Maino, A. (2022). Can unconventional monetary policy contribute to climate action? *Swiss finance institute research paper, 22–35.*

European Central Bank. (2011). *The monetary policy of the ECB* (3rd ed.). https://www. ecb.europa.eu/pub/pdf/other/monetarypolicy2011en.pdf

European Central Bank. (2020). *ECB to accept sustainability-linked bonds as collateral.* https://www.ecb.europa.eu/press/pr/date/2020/html/ecb.pr200922~482e4a5a90. en.html

European Central Bank. (2021a). *Detailed roadmap of climate change-related actions.*

European Central Bank. (2021b). *ECB listens – Summary report of the ECB listens portal responses.* https://www.ecb.europa.eu/home/search/review/html/ecb. strategyreview002.en.html

European Central Bank. (2021c). *ECB presents action plan to include climate change considerations in its monetary policy strategy.* https://www.ecb.europa.eu/press/pr/ date/2021/html/ecb.pr210708_1~f104919225.en.html

European Central Bank. (2022a). *2022 climate risk stress test.* Publications Office. https://data.europa.eu/doi/10.2866/97350

European Central Bank. (2022b). *ECB takes further steps to incorporate climate change into its monetary policy operations.* https://www.ecb.europa.eu/press/pr/date/2022/ html/ecb.pr220704~4f48a72462.en.html

European Central Bank. (2023, February 23). *Foreign reserves and own funds.* https:// www.ecb.europa.eu/ecb/tasks/reserves/html/index.en.html

European Central Bank. (2024a). *Climate and nature plan 2024-2025 at a glance.* https:// www.ecb.europa.eu/ecb/climate/our-climate-and-nature-plan/html/index.en.html

European Central Bank. (2024b). *Climate-related financial disclosures of Eurosystem assets held for monetary policy purposes and of the ECB's foreign reserves.* Publications Office. https://www.ecb.europa.eu/ecb/climate/climate-related-financial-disclosures/ shared/pdf/ecb.crfd2024_MPPs.en.pdf?25ad31ac0ac528fac9ef179d5aefa0ec

European Central Bank. (2024c). *Eurosystem and ECB portfolios steadily decarbonising, climate-related disclosures show.* https://www.ecb.europa.eu/press/pr/date/2024/html/ ecb.pr240625~94de5e0780.en.html

European Parliament. (2020). *European Parliament resolution of 12 February 2020 on the European Central Bank Annual Report for 2018 (2019/2129(INI)).* https://www. europarl.europa.eu/doceo/document/TA-9-2020-0034_EN.pdf

Fender, I., McMorrow, M., Sahakyan, V., & Zulaica, O. (2020). *Reserve management and sustainability: The case for green bonds?, BIS Working Papers No. 849.*

Ferrari, A., & Landi, V. N. (2023). Whatever it takes to save the planet? Central banks and unconventional green policy. *Macroeconomic Dynamics, 28*(2), 299–324.

Fisher, P., & Murphy, D. (2021). *Greening the Central Bank balance sheet – or not?, King's Business School Working Paper No. 2021/6.*

Germann, M., Kusmierczyk, P., & Puyo, C. (2023). Results of the 2022 climate risk stress test of the Eurosystem balance sheet. *Economic Bulletin, 2/2023*, 79–82.

Giovanardi, F., Kaldorf, M., Radke, L., & Wicknig, F. (2023). The preferential treatment of green bonds. *Review of Economic Dynamics, 51*, 657–676.

Giraud, G., Nicol, C., Benaiteau, S., Bonaventure, M., Chaaoub, V., Jego, M., Tsanga, A., Mitteau, G., Philippot, L., & Poidatz, A. (2021). *Fossil assets: The new subprimes? How funding the climate crisis can lead to a financial crisis-June 2021*, https://reclaimfinance.org/site/wp-content/uploads/2021/06/Report-Fossil-Assets-the-new-subprimes.pdf

Goodhart, C. A. E. (2011). The changing role of central banks. *Financial History Review, 18*(2), 135–154.

Grauwe, P. D. (2019). Green money without inflation. *Vierteljahrshefte Zur Wirtschaftsforschung, 88*(2), 51–54.

Grolleau, G., & Weber, C. (2024). The effect of inflation on CO_2 emissions: An analysis over the period 1970–2020. *Ecological Economics, 217*, 108029.

Hansen, L. P. (2022). Central banking challenges posed by uncertain climate change and natural disasters. *Journal of Monetary Economics, 125*, 1–15.

Hensel, J., Mangiante, G., & Moretti, L. (2024). Carbon pricing and inflation expectations: Evidence from France. *Journal of Monetary Economics*, 103593. https://www.sciencedirect.com/science/article/pii/S0304393224000461

Hilmi, N., Djoundourian, S., Shahin, W., & Safa, A. (2022). Does the ECB policy of quantitative easing impact environmental policy objectives? *Journal of Economic Policy Reform, 25*(3), 259–271.

Honohan, P. (2019). Should monetary policy take inequality and climate change into account? *Peterson Institute for International Economics Working Paper*, 18–18.

Hörber, T., & Weber, G. (2022). *The European environmental conscience in EU politics: A developing ideology*. Routledge.

Howard, P. H., & Sterner, T. (2017). Few and not so far between: A meta-analysis of climate damage estimates. *Environmental and Resource Economics, 68*(1), 197–225. https://doi.org/10.1007/s10640-017-0166-z

Knot, K. (2021, February 17). *Getting the Green Deal done – how to mobilize sustainable finance* [Keynote address at an open event organized by Bruegel, 11 February 2021.]. https://www.bis.org/review/r210217d.htm

Konradt, M., McGregor, T., & Toscani, M. F. G. (2024). Carbon prices and inflation in the Euro area. International Monetary Fund. https://ideas.repec.org/p/imf/imfwpa/2024-031.html

Macaire, C., & Naef, A. (2023). Greening monetary policy: Evidence from the People's Bank of China. *Climate Policy, 23*(1), 138–149.

Matikainen, S., Campiglio, E., & Zenghelis, D. (2017). The climate impact of quantitative easing. *Policy Paper, Grantham Research Institute on Climate Change and the Environment, London School of Economics and Political Science, 36*.

McGlade, C., & Ekins, P. (2015). The geographical distribution of fossil fuels unused when limiting global warming to 2 C. *Nature, 517*(7533), 187–190.

Nishigaki, H. (2023). The impact of rising EU allowance prices on core inflation in the Eurozone. *Economic Affairs, 43*(2), 245–264. https://doi.org/10.1111/ecaf.12588

Nordhaus, W. (2018). Projections and uncertainties about climate change in an era of minimal climate policies. *American Economic Journal: Economic Policy, 10*(3), 333–360.

Papoutsi, M., Piazzesi, M., & Schneider, M. (2021). How unconventional is green monetary policy. *Unpublished Manuscript*, 2022.

Reghezza, A., Altunbas, Y., Marques-Ibanez, D., d'Acri, C. R., & Spaggiari, M. (2022). Do banks fuel climate change? *Journal of Financial Stability, 62*, 101049.

Reinders, H. J., Schoenmaker, D., & van Dijk, M. (2023). A finance approach to climate stress testing. *Journal of International Money and Finance, 131*, 102797.

Rennert, K., Errickson, F., Prest, B. C., Rennels, L., Newell, R. G., Pizer, W., Kingdon, C., Wingenroth, J., Cooke, R., & Parthum, B. (2022). Comprehensive evidence implies a higher social cost of CO_2. *Nature, 610*(7933), 687–692.

Rusnok, J. (2021). *The changing world of central banking*. https://www.cnb.cz/en/public/media-service/speeches-conferences-seminars/presentations-and-speeches/The-Changing-World-of-Central-Banking/

Salin, M., Svartzman, R., Biermann, L., Concordet, R., Grisey, L., & Le Calvar, E. (2021). Biodiversity loss and financial stability: A new frontier for central banks and financial supervisors? *Bulletin de La Banque de France, 237*, https://www.banque-france.fr/system/files/2023-01/bdf237-7_biodiversite_vf.pdf

Schnabel, I. (2021a). Climate change and monetary policy. *Finance & Development, 58*(3), 53–55.

Schnabel, I. (2021b). From market neutrality to market efficiency. *International Monetary Review, 8*(3), 69.

Schnabel, I. (2023, January 10). *Monetary policy tightening and the green transition* [Speech at the International Symposium on Central Bank Independence, Sveriges Riksbank, Stockholm]. https://www.ecb.europa.eu/press/key/date/2023/html/ecb.sp230110~21c89bef1b.en.html

Schoenmaker, D. (2021). Greening monetary policy. *Climate Policy, 21*(4), 581–592.

Siedenbiedel, C. (2021, March 13). Der Streit um die grüne Geldpolitik. *FAZ.NET*. https://www.faz.net/aktuell/finanzen/der-streit-um-die-gruene-geldpolitik-17242817.html

Sveriges Riksbank. (2020). *Asset management and sustainability*. https://www.riksbank.se/en-gb/about-the-riksbank/the-riksbanks-work-on-sustainability/asset-management-and-sustainability/

Sveriges Riksbank. (2022). *The Riksbank's work on sustainability*. https://www.riksbank.se/en-gb/about-the-riksbank/the-riksbanks-work-on-sustainability/

Taleb, N. N. (2007). *The black swan: The impact of the highly improbable* (Vol. 2). Random House.

Tucker, P. (2018). *Unelected power: The quest for legitimacy in central banking and the regulatory state*. Princeton University Press.

Various. (2019). *The ECB must act now on climate change* [Open letter to Christine Lagarde].

Volz, U. (2017). On the role of central banks in enhancing green finance. *UN Environment Inquiry Working Paper 17/01*.

Weber, C. S. (2014). ECB: Credibility at risk? *Applied Economics Quarterly, 60*(2), 123–143. https://doi.org/10.3790/aeq.60.2.123

Weber, C. S., & Forschner, B. (2014). ECB: Independence at risk? *Intereconomics, 49*(1), 45–50.

Woolley, J. (2024, January 9). *The ECB is wrong. Green dual interest rates are possible – And necessary*. Green Central Banking. https://greencentralbanking.com/2024/01/09/green-dual-interest-rates-ecb-emmanuel-macron/

Part III

The Impact of Innovation and Knowledge Transfer on Sustainability in the European Union

8 Enabling Sustainability

The Role of Knowledge Flows in Advancing Sustainable Technologies

Naciba Chassagnon-Haned and Amina Hamani

Introduction

In the face of global challenges such as climate change, resource depletion, and environmental degradation, the imperative for sustainable solutions has become increasingly urgent (Rockström et al. 2009). According to Nidumolu et al. (2015), the transition to sustainable development has been recognised as a key challenge for societies worldwide. Sustainable technologies, characterised by their ability to meet current needs without compromising the ability of future generations to meet their own, are at the forefront of addressing these challenges (UNESCO, 2005). Innovation, particularly in the form of technological advancements, plays a pivotal role in addressing sustainability challenges. From renewable energy sources to eco-friendly manufacturing processes, sustainable technologies offer innovative approaches to tackle environmental issues while promoting economic growth and social well-being.

Sustainable technologies refer to the innovative applications of science and engineering that aim to minimise environmental impact, conserve resources, and promote long-term ecological balance while meeting human needs (UNEP, 2018). These technologies prioritise the reduction of carbon emissions, the minimisation of waste generation, and the optimisation of resource utilisation throughout their life cycles. Examples of sustainable technologies include renewable energy sources like solar and wind power, energy-efficient appliances and buildings, waste reduction and recycling technologies, and eco-friendly transportation solutions such as electric vehicles. Europe's proactive stance, exemplified by the EU's ban on the sale of new petrol and diesel cars from 2035, underscores its commitment to promoting eco-friendly transportation and accelerating the transition to sustainable technologies.

However, the path to developing and implementing such technologies is fraught with complexities and challenges. Sustainable technology development often involves addressing complex challenges related to resource scarcity, pollution, and climate change. In this respect, knowledge plays a pivotal role in the development, adoption, and diffusion of sustainable technologies (Hermundsdottir and Aspelund 2021). It serves as the foundation upon which innovations are built,

DOI: 10.4324/9781032656359-12

driving progress and enabling continuous improvement in sustainability practices (Evangelista and Durst 2015; Chopra & Meindl 2021). For example, innovation and research efforts fuelled by knowledge contribute to the creation of new sustainable technologies, improving efficiency, performance, and environmental impact (Jaeger-Waldau 2018). In addition, education and awareness about sustainability issues are essential for promoting the adoption of sustainable technologies (Lutsey and Sperling 2008). Educated individuals and communities are more likely to embrace eco-friendly practices and support sustainable initiatives, driving demand for sustainable products and services (Stern 2000).

Furthermore, academics highlight the importance of collaboration among stakeholders, facilitated by knowledge exchange, which accelerates innovation and scales up sustainable solutions. Sustainable technology development requires collaboration across various disciplines, including engineering, environmental science, economics, and policy. Knowledge facilitates communication and collaboration among experts from different fields, enabling them to work together towards common goals.

Moreover, knowledge of scientific principles and technological capabilities informs the development of policies and regulations aimed at promoting sustainability. Policymakers rely on evidence-based information to design laws and incentives that encourage the adoption of sustainable technologies. Informed policies and regulations, such as the Strategic Technologies for Europe Platform and the Net-Zero Industry Act, incentivise investment in sustainable technologies and drive industry-wide changes towards more sustainable practices. Additionally, initiatives like Horizon Europe and the European Chips Act further support innovation and research in sustainable technologies.

Data protection and handling also play crucial roles, with regulations such as the General Data Protection Regulation (GDPR), the Digital Markets Act (DMA), and the Digital Services Act (DSA) ensuring that innovation and research activities respect privacy and data security.

Furthermore, institutions like the European Innovation Council (EIC), the Joint Research Centre (JRC), and the EU Competence Centre on Technology Transfer support sustainable technology development through various programmes. The JRC's Competence Centre on Technology Transfer, for instance, supports projects in Capacity Building, Financing, Science and Technology Parks, and Innovation Ecosystems. Other key institutions include the European Institute of Technology (EIT) and the European Research Area (ERA). Finally, continuous improvement of sustainable technologies is enabled by knowledge-driven feedback loops, data analysis, and lessons learned from real-world implementation, leading to more effective, efficient, and sustainable solutions (Reap et al. 2008). Knowledge allows for the optimisation of sustainable technologies by fine-tuning their design, materials, and processes. Through continuous learning and experimentation, developers can enhance the performance and cost-effectiveness of these technologies.

The aim of this chapter is to explore the importance of knowledge flows exploitation and exploration for more sustainable technologies, as well as the role of the policy mix in this process. We will begin by reviewing the literature on knowledge flows and their significance in sustainable technology development. Then, we will provide concrete examples of how knowledge flow and exchange drive sustainable technologies and innovation in various sectors. Finally, we will offer recommendations for policymakers and stakeholders to leverage knowledge flows effectively in advancing sustainable technologies.

Literature Review

Defining Sustainable Technologies

Sustainable technologies encompass innovative solutions devised to address societal needs while minimising ecological footprints and optimising resource utilisation. These technologies play a pivotal role in sustainable development by seeking to reconcile economic growth, social equity, and environmental stewardship. Notable examples include Solar Photovoltaic Systems and Energy-Efficient Building Materials, which serve as renewable energy sources aimed at mitigating greenhouse gas emissions and countering climate change. Empirical investigations, such as those conducted by Xiong et al. (2020), substantiate the efficacy of these technologies in mitigating environmental degradation and curbing energy consumption.

The imperative for sustainability serves as a central driver of innovation across diverse sectors, stimulating the emergence of sustainable value creation strategies that address interconnected environmental, social, and economic challenges. However, it is essential to address potential unintended consequences of technological solutions (Mulder, Ferrer, and Lente 2017). While a technology may solve one problem, it could create or exacerbate others. Therefore, thorough impact assessments are necessary to evaluate the social, environmental, and economic implications of adopting the technology. The effects of a technology depend not only on its inherent characteristics but also on how it is perceived, used, and integrated into society. Therefore, a holistic approach to technology assessment is essential, considering both short-term and long-term consequences.

This evolution towards sustainability-focused innovation is observable today in various industries, including energy, transportation, agriculture, and manufacturing, each witnessing distinctive advancements. In the energy sector, the transition towards renewable energy sources like solar, wind, and hydroelectric power exemplifies how sustainability considerations are fostering innovation. Noteworthy progress in renewable energy technologies has been facilitated by intensified research endeavours and investment, resulting in enhanced efficiency and cost-effectiveness, as documented by Bhattacharyya (2011). Furthermore, the development of energy storage solutions, such as advanced batteries and grid-scale storage systems, assumes critical importance for the seamless integration of renewable energy into power grids, ensuring energy stability and reliability.

Reconceptualising traditional business models to incorporate sustainability principles represents another facet of innovation, generating new value streams while mitigating adverse societal and environmental impacts (Schaltegger, Lüdeke-Freund, and Hansen 2016). Instances of innovative business models integrating circular economy principles, such as product-as-a-service, remanufacturing, and waste-to-value strategies, exemplify how economic sustainability can be pursued alongside societal and environmental benefits (Geissdoerfer et al., 2017; Boons et al., 2013).

In summary, sustainability-driven innovation is fundamentally reshaping organisational operations, competitive landscapes, and value creation mechanisms, with far-reaching implications for environmental preservation, social welfare, and economic prosperity. By embedding sustainability principles into their innovation strategies and business practices, companies can ensure long-term viability and contribute to a more sustainable and resilient global future. However, to ensure that technologies effectively contribute to sustainable development, it is crucial to establish.

Understanding Knowledge Flows

Knowledge flows refer to the movement and exchange of knowledge, information, and expertise among individuals, organisations, and systems (Chopra & Meindl 2021). These flows are crucial for innovation, problem-solving, and decision-making processes (Evangelista and Durst 2015; Argote and Ingram 2000; Chopra & Meindl 2021). The literature often distinguishes two different types of knowledge flows, each with its own characteristics and implications (Krogh, Ichijo, and Nonaka 2000). Firstly, tacit knowledge, which refers to personal, unspoken knowledge that is difficult to articulate or codify. It is often based on personal experiences, skills, and intuition. Tacit knowledge is shared through direct interaction, observation, and hands-on experience. Secondly, explicit knowledge, on the other hand, is formalised, codified, and easily transferable. It includes documented information, data, procedures, and theories that can be communicated through written documents, manuals, databases, or digital platforms. Examples of explicit knowledge include scientific papers, technical manuals, and instructional videos.

The systematic exploration of these types of knowledge flows entails a nuanced investigation into the intricacies of the processes by which knowledge is generated, disseminated, transferred, and utilised across diverse organisational, sectoral, and geographical contexts. This scholarly pursuit aims to elucidate the intricate mechanisms and dynamics underlying knowledge creation, documentation, and sharing, which in turn serve to stimulate innovation, advance educational endeavours, and address multifaceted challenges (Nonaka and Takeuchi, 1995). Instances of knowledge flows manifest in various forms, including structured collaborations, technology exchange agreements, informal networking arrangements, and serendipitous knowledge spillovers, all of which play fundamental roles in the functioning of the knowledge economy (Argote & Ingram, 2000).

Empirical research has underscored the significance of collaborative research initiatives that involve partnerships among academic institutions, corporate entities, and governmental organisations in facilitating the reciprocal exchange of expertise, resources, and information. This collaborative synergy is pivotal in engendering innovative ideas and practices across diverse domains such as biotechnology, renewable energy, and information technology (Etzkowitz and Leydesdorff, 2000).

Additionally, the role of technology transfer and licensing agreements in facilitating knowledge flows, particularly between academic institutions, research organisations, and the industrial sector, is critical for the commercialisation of research findings and the transfer of intellectual property rights. These mechanisms are essential for introducing novel products, methodologies, and services into the market, thereby fostering economic development, and enhancing industrial competitiveness (Mowery et al., 2001). Informal networks and communities of practice also serve as crucial conduits for facilitating knowledge flows, where tacit knowledge, best practices, and experiential learning are exchanged among individuals and organisations. These informal platforms significantly augment peer-to-peer learning, collaborative projects, and the widespread dissemination of knowledge, transcending geographical and organisational boundaries (Brown and Duguid, 2001).

In the contemporary knowledge-driven economy, open innovation frameworks and crowdsourcing initiatives have emerged as potent mechanisms for harnessing the collective intelligence of a diverse array of external stakeholders, including customers, suppliers, and independent researchers. These platforms facilitate the co-creation of knowledge and the identification of innovative solutions to complex challenges, thereby amplifying the societal and economic impact of knowledge flows (Chesbrough, 2003; Howe, 2006).

Empirical Evidence and Case Studies

Renewable Energy Technologies

Existing literature suggests that knowledge flow plays a crucial role in driving the development and adoption of sustainable technologies through research advancement, educational enrichment, and the promotion of collaboration among stakeholders. For instance, the German Energiewende initiative emphasises the significant role of knowledge flow in advancing sustainable energy transitions. This initiative demonstrates how active engagement among researchers, policymakers, and industry stakeholders facilitates knowledge exchange and co-creation of solutions (Gailing and Röhring 2016).

Several analogous initiatives vividly demonstrate the significant influence of disseminating knowledge emanating from both the research community and diverse stakeholders across various fields and regions globally. The second example that is interesting to discuss in this paper is that of the German Energiewende initiative, which sparks significant interest and debate within the academic community. As one of the most ambitious national energy transition initiatives globally, it presents a unique opportunity to understand the importance of knowledge flow in transitions to renewable energies. The German Energiewende initiative aims to shift towards renewable energy sources and phase out nuclear power by 2022. Despite progress, the Energiewende faces criticism and scepticism, particularly regarding its ability to meet ambitious environmental and energy reduction targets while phasing out nuclear generation and dealing with the dominance of the coal industry. According to recent research, overcoming these challenges and ensuring a successful transition to a sustainable energy future necessitates continued knowledge sharing and stakeholder collaboration (Paul and Rather,2018). Further researchers suggest that international collaboration facilitated by knowledge exchange is essential for addressing global climate and energy challenges, underscoring the significance of knowledge as a catalyst for innovation, collaboration, and informed decision-making in advancing sustainable technology within the context of the German Energiewende (Gailing and Röhring 2016).

Sustainable Agriculture and Food Systems

We observe that the exchange of knowledge has been pivotal in driving the adoption and advancement of sustainable agricultural technologies in many countries, with precision farming, agroforestry, conservation tillage, and integrated pest management serving as notable examples (Wolfert et al., 2017). For instance, in Kenya, initiatives led by organisations such as the World Agroforestry Centre (ICRAF) have facilitated the dissemination of knowledge about agroforestry practices. Through research, training programmes, and extension services, farmers have gained insights into the benefits and techniques of integrating trees into their farming systems. This knowledge exchange has empowered Kenyan farmers to implement agroforestry systems tailored to their specific agroecological contexts, addressing challenges such as soil degradation, water scarcity, and food insecurity. In Brazil, the flow of knowledge has contributed to the widespread adoption of agroforestry systems in the Amazon rainforest region. Research institutions, NGOs, and government agencies have collaborated to promote sustainable land use practices that combine agriculture with tree cover. By sharing information about the ecological and economic benefits of agroforestry, stakeholders have encouraged farmers to implement these systems as a means of protecting biodiversity, mitigating deforestation, and enhancing rural livelihoods.

Similarly, in the United States, knowledge exchange efforts have facilitated the adoption of agroforestry practices for various purposes, including riparian buffer restoration and windbreak establishment. The USDA National Agroforestry Center, along with other organisations, has provided technical assistance, educational resources, and funding opportunities to support agroforestry initiatives nationwide. By sharing research findings, case studies, and practical guidance, stakeholders have enabled American farmers to implement agroforestry practices that enhance soil and water conservation, biodiversity conservation, and economic resilience. Overall, the flow of knowledge has been instrumental in driving the advancement of sustainable agricultural technologies within these countries by providing farmers with the information, tools, and support needed to adopt innovative practices that improve productivity, profitability, and environmental stewardship.

Green Building and Urban Infrastructure

Knowledge exchange has been pivotal in driving the adoption and advancement of sustainable agricultural technologies in many countries. For example, in Kenya, initiatives led by organisations such as the World Agroforestry Centre (ICRAF) have facilitated the dissemination of knowledge about agroforestry practices. This knowledge exchange has empowered Kenyan farmers to implement agroforestry systems tailored to their specific agroecological contexts (Wolfert et al. 2017).

Singapore is innovating in water and waste management technologies, which include advanced wastewater treatment plants, rainwater harvesting systems, advanced waste-to-energy facilities, waste sorting technologies, and innovative recycling processes. These sustainable technologies not only contribute to reducing carbon emissions, conserving resources, and enhancing overall quality of life for residents but also play a crucial role in supporting Singapore's goals of creating a resilient, environmentally friendly urban landscape(Radhakrishnan et al. 2019). Through knowledge exchange and collaboration, stakeholders in Singapore are driving innovation and fostering the adoption of these technologies to address current environmental challenges and create a sustainable future for generations to come.

Circular Economy and Resource Management

The role of knowledge in advancing sustainable technologies within the Circular Economy and Resource Management is widely explored by researchers. We choose to highlight this role in the case of Sweden's Circular Economy Strategy, where knowledge flow plays a pivotal role in driving the adoption and advancement of sustainable technology. This strategy focuses on transitioning towards a more sustainable and circular economy by minimising waste generation, maximising resource efficiency, and reducing environmental impact. Research and Development (R&D) programmes in Sweden play a vital role in driving innovation for the circular economy. Through collaborative efforts involving academia, industry, and government agencies, innovative technologies for waste management, recycling, and renewable energy are developed.

Sweden consistently ranks highly on the European Innovation Scoreboard, published by the European Commission, reflecting its commitment to R&D investment. The business sector leads in R&D investment, with significant contributions, while government agencies like Vinnova play a crucial role in promoting and funding research projects across various fields.

Overall, these case studies emphasise the interconnectedness of knowledge exchange, innovation, and policy influence in advancing sustainable technology and achieving broader sustainability goals. Table 8.1 compares the impact of knowledge flow on sustainable technology advancement across different sectors: Renewable Energy Technologies, Agroforestry, Green Building and Urban Infrastructure, and Circular Economy and Resource Management. Each sector is analysed through case studies highlighting technologies, drivers, challenges, and the role of knowledge exchange. Key insights include the importance of collaboration, the adoption of innovative practices, and the role of capacity building in driving sustainability goals.

Table 8.1 The Impact of Knowledge Flow on Sustainable Technology Advancement Across Different Sectors

Case Study	German Energiewende	Agroforestry in Kenya, Brazil, and the United States	Singapore's Sustainable Urban Planning Initiatives	Sweden's Circular Economy Strategy
Sector	Renewable Energy Technologies	Sustainable Agriculture	Green Building and Urban Infrastructure	Circular Economy and Resource Management
Technologies	Solar PV systems, wind turbines, energy storage systems, smart grids	Precision farming, conservation tillage, agroforestry, integrated pest management	Smart grid systems, energy-efficient lighting, intelligent transportation systems	Waste management technologies, recycling technologies, renewable energy technologies, circular design
Knowledge Exchange and Capacity Building	Active engagement of interconnected authors' networks with stakeholders, policy formulation, decision-making processes	Research, training programmes, extension services, technical assistance	Green Building Masterplan, collaboration among stakeholders	Research and Development (R&D) programmes, collaboration, capacity building
Ecosystem and Collaboration	Knowledge sharing among research, policymakers, industry, communities, international collaboration	Collaboration between research institutions, NGOs, government agencies, international organisations	Partnerships between governmental agencies, research institutions, international organisations	Partnerships between academia, industry, government agencies, international organisations

Challenges and Opportunities

Barriers to Knowledge Flows

Barriers and Enablers to knowledge flow

Barriers to knowledge flows pose formidable obstacles to the efficient dissemination, transfer, and utilisation of knowledge within and across organisational, sectoral, and regional contexts. These barriers arise from various sources, including organisational structures, technological constraints, cultural disparities, and institutional frameworks. Understanding these impediments is essential for devising strategies to surmount them and promote effective knowledge exchange and collaboration. The subsequent sections delineate key barriers to knowledge flows, supported by scholarly literature:

Organisational silos manifest as compartmentalisation within entities, fostering rigid divisions among departments, teams, or units. Such compartmentalisation can hinder the cross-sectional flow of knowledge, leading to redundant work, communication breakdowns, and restricted information and expertise exchange. Studies underscore the adverse effects of organisational silos on knowledge sharing and collaboration, elucidating how they contribute to operational inefficiencies and missed opportunities for innovation (Cummings, 2004).

Trust and collaborative dynamics are foundational to successful knowledge exchange. A deficit of trust between individuals or entities can engender reluctance to share knowledge, driven by concerns about reputation, competitive disadvantage, or expectations of reciprocation. Additionally, inadequate mechanisms for collaboration can diminish the efficacy of communication and coordination. Trust and social capital are imperative for enabling knowledge exchange and collaborative endeavours within and across organisational networks (Nahapiet and Ghoshal, 1998).

Technological barriers, such as outdated or incompatible IT systems, limited access to digital tools, and cybersecurity issues, can impede knowledge flow. Ill-conceived technological platforms or a dearth of digital literacy can challenge the effective sharing, accessing, or application of knowledge. The challenges posed by technological barriers in knowledge sharing within virtual teams underscore the importance of user-friendly interfaces and secure communication channels (Yang et al., 2019).

Cultural differences, encompassing language barriers, communication styles, and varying attitudes toward knowledge sharing, can obstruct cross-cultural collaboration and knowledge exchange. Disparities in values, norms, and social structures can impact individuals' willingness to openly share knowledge. Hansen et al. (2005) investigate the influence of cultural diversity on knowledge transfer within multinational organisations, advocating for cultural sensitivity and cross-cultural training to overcome communicative hurdles.

The absence of suitable recognition, incentives, and rewards for knowledge sharing and collaboration can dissuade individuals from contributing their expertise. Without adequate motivational systems, there may be a propensity to prioritise personal objectives over collective goals, leading to knowledge hoarding or reluctance to engage in collaborative endeavours. In this framework, the pivotal role of incentives and reward systems in promoting knowledge sharing within organisations underscores the necessity of aligning individual incentives with organisational objectives (Szulanski, 1996).

Addressing these barriers necessitates a comprehensive approach that fosters a culture of openness and trust, invests in appropriate technological infrastructure, promotes interdisciplinary collaboration, and establishes incentives and recognition for knowledge-sharing behaviours (Alavi & Leidner, 2001). By tackling these obstacles, organisations and policymakers can create conditions conducive to seamless knowledge flow, thereby stimulating innovation, learning, and enhanced organisational performance.

Facilitators for efficient knowledge transfer are integral to enabling the unhindered exchange, dissemination, and application of knowledge both within and across organisational boundaries. These facilitators can be diverse, encompassing aspects such as organisational culture, leadership endorsement, communication frameworks, and technological infrastructure. Recognising and harnessing these facilitators is crucial for the advancement of successful knowledge transfer endeavours.

The following segments outline key facilitators for effective knowledge transfer

An organisational culture that values continuous learning, collaboration, and the open sharing of knowledge establishes a fertile ground for effective knowledge transfer. An environment where employees are motivated and supported to exchange expertise and learn reciprocally enhances the fluidity of knowledge across the organisational spectrum. Nonaka and Takeuchi (1995) elaborate on the "knowledge-creating" company model, which nurtures a culture of knowledge sharing and innovation, underscoring the pivotal role of leadership, a unified vision, and organisational norms in propelling knowledge creation and transfer.

Leadership plays a crucial role in advocating for knowledge transfer initiatives and cultivating an ecosystem supportive of knowledge exchange. Leadership commitment, resource allocation, and the establishment of explicit knowledge-sharing expectations significantly influence employee participation in knowledge transfer activities. Top management's contribution to knowledge transfer facilitation within organisations highlights the critical nature of leadership dedication, communicative clarity, and the alignment of motivational incentives (Zahra and George, 2002).

Robust communication frameworks, including routine meetings, knowledge-sharing platforms, and professional communities, serve as conduits for the exchange of ideas, insights, and best practices. Transparent communication cultivates trust and collaborative synergy among individuals and teams.

Communication on processes and technologies in knowledge transfer accentuates the value of informal networks, narrative techniques, and social interactions in the circulation of tacit knowledge (Davenport and Prusak, 1998).

Training and development programmes focused on augmenting employees' knowledge-sharing competencies, communicative adeptness, and interdisciplinary collaboration skills significantly enhance their capacity for effective knowledge transfer. Opportunities for skill enhancement and learning reinforce a knowledge-sharing ethos. The efficacy of training programmes and socialisation strategies in fostering knowledge transfer emphasises the roles of experiential learning, feedback mechanisms, and community engagement practices (Argote and Ingram, 2000).

By capitalising on these facilitators, organisations can cultivate an ecosystem that nurtures effective knowledge transfer, thereby contributing to enhanced innovation, problem-solving capabilities, and overall organisational performance.

Enhancing Knowledge Sharing Mechanisms

The enhancement of knowledge-sharing mechanisms within organisational contexts can be elucidated through a synthesis of various theoretical frameworks and empirical evidence, delineating the multifaceted strategies, processes, and technological interventions that facilitate such dynamics. This discourse integrates pivotal academic discourses to underscore the mechanisms and their underpinning rationales:

Social Capital Theory posits the foundational role of social networks, trust, and interpersonal connections in catalysing knowledge exchange within organisations. It suggests that the cultivation of robust social relationships and the establishment of a reciprocal ethos significantly contribute to the efficacy of knowledge-sharing practices. Empirical inquiries within this domain explore the nuanced impact of social capital components on knowledge dissemination behaviours across diverse organisational landscapes, focusing on the influence of trust, network structures, and collective identity in enhancing collaborative engagements among personnel (e.g., Nahapiet & Ghoshal, 1998).

Communities of Practice represent informal conglomerations of individuals unified by common interests or expertise, engaging in collaborative knowledge exchange and problem-solving endeavours. These entities serve as conduits for organisational learning and innovation by fostering a vibrant exchange of knowledge and expertise among participants. Research in this area scrutinises elements such as leadership dynamics and organisational support mechanisms that underlie the operational success of these communities (e.g., Wenger, McDermott, & Snyder, 2002).

Knowledge Management Systems (KMS) encompass the technological and procedural infrastructure dedicated to the facilitation of knowledge generation, organisation, and dissemination within enterprises. These systems, including document repositories and collaborative platforms, underpin organisational learning processes and informed decision-making. Scholarly investigations in this field delve into the design, deployment, and impacts of KMS on knowledge-sharing behaviours and outcomes, with a particular focus on the influence of design aesthetics, user interface, and organisational culture on system efficacy (e.g., Alavi & Leidner, 2001).

Incentive and Recognition Mechanisms are critical in incentivising knowledge-sharing behaviours among organisational members. By instituting reward structures and acknowledgement schemes for knowledge contributions, entities can stimulate a culture of knowledge dissemination. Studies in this domain assess the ramifications of various incentive and recognition frameworks on knowledge-sharing dynamics and outcomes, evaluating factors such as perceived fairness and congruence with overarching organisational objectives (e.g., Wasko & Faraj, 2005).

Through the integration of insights from these theoretical and empirical lenses, organisations are poised to architect and implement nuanced strategies that bolster knowledge sharing among employees, thereby nurturing a culture of enhanced collaboration, innovation, and overall organisational efficacy.

Leveraging Emerging Technologies for Knowledge Exchange

In the domain of leveraging emerging technologies for knowledge exchange, a synthesis of theoretical frameworks and empirical evidence provides a comprehensive understanding of the dynamics at play. This integrated analysis encompasses several key theories and their practical applications:

Rooted in the premise that individuals acquire knowledge through observation, imitation, and social interaction, Social Learning Theory underscores the potential of emerging technologies like social media, online forums, and collaborative platforms to facilitate this process. These digital environments offer rich opportunities for individuals to share and assimilate knowledge, expertise, and experiences with their peers, thereby enhancing collective learning. Investigative efforts within this theoretical boundary focus on delineating the social learning mechanisms in digital contexts and assessing the influence of technological advancements on knowledge sharing practices and outcomes. Empirical studies scrutinise variables such as social presence, trust, and established community norms to discern their effects on knowledge dissemination behaviours (Lin et al., 2018).

CoPs are conceptualised as collectives where individuals with shared interests or professional backgrounds engage in communal learning and knowledge exchange. Emerging technologies offer vital support structures for CoPs, enabling them to share insights, best practices, and expertise through online communities, virtual collaboration platforms, and digital knowledge repositories.

Empirical investigations within the CoP framework delve into how emerging technologies bolster knowledge sharing and collaborative endeavours within and across such communities. Research focuses on identifying elements critical to CoP efficacy, including leadership dynamics, member participation, and technological affordances, to elucidate their contributions to organisational learning and innovation (Wenger et al., 2002).

Blockchain technology is heralded for its capacity to offer decentralised, transparent mechanisms for secure information recording, verification, and exchange. In the realm of knowledge exchange, it promises to underpin systems where intellectual property rights, credentials, and transaction records are immutably logged, fostering trustworthy and verifiable sharing environments. Analysis of blockchain's role in knowledge exchange traverse its utility in managing intellectual property, enhancing supply chain visibility, and overseeing digital rights. These studies assess the practicality, scalability, and integration challenges of blockchain-enabled knowledge exchange frameworks across various sectors and scenarios (Swan, 2015).

Conclusion

The urgency of addressing global challenges such as climate change, resource depletion, and environmental degradation has never been clearer. Sustainable technologies offer promising solutions to these challenges, aiming to meet current needs while safeguarding the ability of future generations to meet their own. However, the development and implementation of sustainable technologies are intricate processes, rife with complexities and obstacles.

Throughout this chapter, we have underscored the pivotal role of knowledge in advancing sustainable technologies. Knowledge serves as the bedrock upon which innovations are built, driving progress and enabling continuous improvement in sustainability practices. From the creation of new technologies to the promotion of eco-friendly behaviours, knowledge is the driving force behind sustainable development. We have highlighted different theoretical perspectives on knowledge transfer and management that offer valuable insights into how knowledge flows drive technological innovation. We reveal that Social Network Theory, Absorptive Capacity Theory, and Communities of Practice Theory provide frameworks for understanding knowledge exchange within and among organisations. In sustainable technology development, theories like the Knowledge-Based View and the Innovation Systems Approach emphasise the importance of knowledge creation and diffusion. Additionally, theories such as Social Network Theory and Absorptive Capacity Theory shed light on how knowledge is disseminated and absorbed, while concepts like Social Capital Theory and Collaborative Innovation Networks highlight the role of collaboration and networks in knowledge exchange. Therefore, understanding these theoretical perspectives helps stakeholders develop strategies to foster collaboration, enhance knowledge exchange, and drive sustainable technological innovation.

Moreover, across the diverse domains of renewable energy technologies, sustainable agriculture, green building and urban infrastructure, and circular economy initiatives, several key findings emerge. First, the importance of collaborative networks between researchers, policymakers, industry stakeholders, and communities is vital for driving innovation and informing decision-making processes. Secondly, the role of Digital innovations such as IoT, cloud computing, and AI plays a crucial role in facilitating knowledge exchange, optimising resource utilisation, and driving sustainable technology development. Furthermore, the Impact on Policy and Practice, where Knowledge flow influences policy formulation, technology adoption, and business model innovation, shapes the trajectory of sustainability initiatives at local, national, and global scales. Finally, the need for Capacity building initiatives that empower individuals and organisations to effectively implement sustainable technologies, driving widespread adoption and positive environmental outcomes.

This chapter also discusses the challenges and opportunities in knowledge transfer and sharing within inter-organisational contexts. It demonstrates that challenges are multifaceted, encompassing barriers, facilitators, and the leveraging of emerging technologies. Barriers to knowledge flows, such as organisational silos, trust deficits, technological limitations, cultural disparities, and inadequate recognition systems, impede the efficient dissemination and utilisation of knowledge. Overcoming these barriers requires strategies that foster openness, trust, technological infrastructure, cultural sensitivity, and appropriate incentives. Furthermore, enhancing knowledge-sharing mechanisms involves integrating theoretical frameworks such as Social Capital Theory, Communities of Practice, Knowledge Management Systems, and Incentive Mechanisms. These frameworks emphasise the importance of social networks, collaborative communities, technological infrastructure, and incentives in facilitating knowledge exchange within organisations.

In exploring the importance of knowledge flows in advancing sustainable technologies, it is evident that effective knowledge management is crucial. The insights gleaned from this chapter have significant implications for both research and practice. First, future research endeavours should delve deeper into understanding the intricate dynamics of knowledge flow within and across sectors. This includes investigating how factors such as organisational structures, cultural norms, and technological infrastructures influence the exchange of knowledge and its impact on sustainable technology development. Researchers should explore the potential of emerging technologies such as blockchain, artificial intelligence (AI), and biotechnology in driving sustainable innovation. Understanding how these technologies can facilitate knowledge exchange, optimise resource utilisation, and address sustainability challenges will be critical for informing future research agendas. There is a need to further examine the effectiveness of cross-sectoral collaboration in advancing sustainability goals. Research can focus on identifying successful models of collaboration between different sectors and exploring the factors that facilitate or hinder knowledge exchange and cooperation.

Looking ahead, several avenues for practice emerge. Practitioners should prioritise initiatives aimed at facilitating knowledge exchange among stakeholders. This includes establishing platforms for sharing best practices, organising workshops and seminars, and fostering communities of practice where individuals can collaborate and learn from each other. Given the significant role of digital technologies in driving knowledge exchange and innovation, organisations should invest in robust digital infrastructure. This includes implementing tools and platforms that facilitate data sharing, collaboration, and communication among stakeholders. Collaboration is key to driving sustainable technology development. Practitioners should actively seek out collaborative partnerships with researchers, policymakers, industry stakeholders, and communities to leverage collective expertise and resources in addressing sustainability challenges.

This chapter concludes by proposing different future directions that, through embracing them and continuing to prioritise knowledge exchange, collaboration, and innovation, economies can pave the way towards a more sustainable and resilient future for all. Future research could explore the potential of emerging technologies. For example, researchers can study the potential of blockchain technology in enhancing transparency, traceability, and trust in sustainable supply chains. Researchers can further explore how AI can be used to analyse large datasets and identify patterns and trends related to sustainability. Additionally, the intersection of biotechnology and sustainability offers promising avenues for innovation. Future research could focus on developing bio-based materials, biodegradable plastics, and sustainable agricultural practices that reduce reliance on synthetic inputs and minimise environmental impact.

Moreover, this chapter has noticed a need for more interdisciplinary research that brings together experts from different fields to tackle complex sustainability challenges. Future initiatives could focus on integrating insights from fields such as ecology, economics, sociology, and engineering to develop holistic solutions. Additionally, as collaboration between the public and private sectors can play a crucial role in driving sustainability initiatives, future efforts could explore innovative partnership models that leverage the strengths of both sectors to accelerate the transition towards a more sustainable future. Finally, future research and practice could focus on developing participatory approaches that empower communities to actively participate in decision-making processes and co-create solutions that are tailored to their needs and aspirations.

Bibliography

Alavi, M., & Leidner, D. E. (2001). Knowledge management and knowledge management systems: Conceptual foundations and research issues. *MIS Quarterly*, 25(1), 107–136.

Argote, L., & Ingram, P. (2000). Knowledge transfer: A basis for competitive advantage in firms. *Organizational Behavior and Human Decision Processes, 82*(1), 150–169.

Bansal, P., & Song, H. C. (2017). Similar but not the same: Differentiating corporate sustainability from corporate responsibility. *Academy of Management Annals*, 11(1), 105–149.

Berman, B. (2012). 3-D printing: The new industrial revolution. *Business Horizons*, 55(2), 155–162.

Bhattacharyya, S. C. (2011). Concepts, issues, markets and governance. *Energy Economics*, 978.

Boons, F., Montalvo, C., Quist, J., & Wagner, M. (2013). Sustainable innovation, business models and economic performance: An overview. *Journal of Cleaner Production*, 45, 1–8.

Brown, J. S., & Duguid, P. (2001). Knowledge and organization: A social-practice perspective. *Organization Science*, 12(2), 198–213.

Chesbrough, H. W. (2003). *Open innovation: The new imperative for creating and profiting from technology*. Harvard Business School Press.

Chopra, S., & Meindl, P. (2021). *Supply chain management. Strategy, planning & operation*. 7th edition. Pearson.

Cohen, W. M., & Levinthal, D. A. (1990). Absorptive capacity: A new perspective on learning and innovation. *Administrative Science Quarterly*, 35(1), 128–152.

Cummings, J. N. (2004). Work groups, structural diversity, and knowledge sharing in a global organization. *Management Science*, 50(3), 352–364.

Cummings, J. N., & Teng, B. S. (2003). Transferring R&D knowledge: The key factors affecting knowledge transfer success. *Journal of Engineering and Technology Management*, 20(1–2), 39–68.

Davenport, T. H., & Prusak, L. (1998). *Working knowledge: How organizations manage what they know*. Harvard Business Press.

Engel, J. S. (2009). The role of clustered interfirm networks in innovation and entrepreneurship. *Journal of Management Studies*, 46(2), 265–292.

Etzkowitz, H., & Leydesdorff, L. (2000). The dynamics of innovation: From National Systems and "Mode 2" to a triple helix of university-industry-government relations. *Research Policy*, 29(2), 109–123.

Evangelista, P., & Durst, S. (2015). Knowledge management in environmental sustainability practices of third-party logistics service providers. *Vine*, *45*(4), 509–529.

Fagnant, D. J., & Kockelman, K. (2018). The travel and environmental implications of shared autonomous vehicles, using agent-based model scenarios. *Transportation Research Part C: Emerging Technologies*, 89, 205–224.

Fuss, S., Lamb, W. F., Callaghan, M. W., Hilaire, J., Creutzig, F., Amann, T., … & Minx, J. C. (2018). Negative emissions—Part 2: Costs, potentials and side effects. *Environmental Research Letters*, 13(6), 063002.

Gailing, L., & Röhring, A. (2016). Germany's Energiewende and the spatial reconfiguration of an energy system. *Conceptualizing Germany's Energy Transition: Institutions, Materiality, Power, Space*, 11–20.

Geissdoerfer, M., Savaget, P., Bocken, N. M., & Hultink, E. J. (2017). The circular economy—A new sustainability paradigm? *Journal of Cleaner Production*, 143, 757–768.

Geng, Y., Fu, J., Sarkis, J., Xue, B., & Fujita, T. (2012). Towards a national circular economy indicator system in China: An evaluation and critical analysis. *Journal of Cleaner Production*, 23(1), 216–224.

Grant, R. M. (1996). Toward a knowledge-based theory of the firm. *Strategic Management Journal*, 17(S2), 109–122.

Hansen, M. T., Nohria, N., & Tierney, T. (2005). What's your strategy for managing knowledge. *Knowledge Management: Critical Perspectives on Business and Management*, 77(2), 322.

Hermundsdottir, F., & Aspelund, A. (2021). Sustainability innovations and firm competitiveness: A review. *Journal of Cleaner Production*, *280*, 124715.

Howe, J. (2006). The rise of crowdsourcing. *Wired Magazine*, 14(6), 1–4.

Jaeger-Waldau, A., PV Status Report 2018, EUR 29463 EN, Publications Office of the European Union, Luxembourg, 2018, ISBN 978-92-79-97466-3, doi:10.2760/924363, JRC113626.

Krogh, G., Ichijo, K., & Nonaka, I. (2000). *Enabling knowledge creation: How to unlock the mystery of tacit knowledge and release the power of innovation*. Oxford University Press.

Lin, Y., et al. (2018). Understanding Knowledge Sharing in Virtual Communities: An Integration of Social Capital and Social Cognitive Theories. *Information & Management*, 55(5), 580–590.

Lozano, R. (2015). A holistic perspective on corporate sustainability drivers. *Corporate Social Responsibility and Environmental Management*, 22(1), 32–44.

Lundvall, B.-Å. (1992). *National systems of innovation: Towards a theory of innovation and interactive learning*. Pinter.

Lutsey, N., & Sperling, D. (2008). America's bottom-up climate change mitigation policy. *Energy policy*, 36(2), 673–685.

Mowery, D. C., Nelson, R. R., Sampat, B. N., & Ziedonis, A. A. (2001). The growth of patenting and licensing by U.S. universities: An assessment of the effects of the Bayh-Dole act of 1980. *Research Policy*, 30(1), 99–119.

Mulder, K., Ferrer, D., & Van Lente, H. (2017). *What is sustainable technology?: Perceptions, paradoxes and possibilities*. Routledge.

Nahapiet, J., & Ghoshal, S. (1998). Social capital, intellectual capital, and the organizational advantage. *The Academy of Management Review*, 23(2), 242.

Nidumolu, R., Prahalad, C. K., & Rangaswami, M. R. (2015). Why sustainability is now the key driver of innovation. IEEE Engineering Management Review, 43(2), 85–91.

Nonaka, I., & Takeuchi, H. (1995). The knowledge-creating company: How Japanese companies create the dynamics of innovation. Oxford University Press.

OCDE. (2006). *Infrastructure to 2030: Telecom, land transport, water and electricity*, Éditions OCDE.

Paul, S., & Rather, Z. H. (2018). A pragmatic approach for selecting a suitable wind turbine for a wind farm considering different metrics. *IEEE Transactions on Sustainable Energy*, 9(4), 1648–1658.

Radhakrishnan, M., Kenzhegulova, I., Eloffy, M. G., Ibrahim, W. A., Zevenbergen, C., & Pathirana, A. (2019). Development of context specific sustainability criteria for selection of plant species for green urban infrastructure: The case of Singapore. *Sustainable Production and Consumption*, 20, 316–325.

Reap, J., Roman, F., Duncan, S., & Bras, B. (2008). A survey of unresolved problems in life cycle assessment: Part 1: goal and scope and inventory analysis. *The International Journal of Life Cycle Assessment*, 13, 290–300.

Rockström, J., Steffen, W., Noone, K., Persson, Å., Chapin, F. S., Lambin, E. F., ... & Foley, J. A. (2009). A safe operating space for humanity. *Nature*, 461(7263), 472–475.

Rogers, E. M. (2003). *Diffusion of innovations* (5th ed.). Free Press.

Rosenbusch, N., Gusenbauer, M., Hatak, I., Fink, M., & Meyer, K. E. (2019). Innovation offshoring, institutional context and innovation performance: A meta-analysis. *Journal of Management Studies*, 56(1), 203–233.

Schaltegger, S., Lüdeke-Freund, F., & Hansen, E. G. (2016). Business models for sustainability: A co-evolutionary analysis of sustainable entrepreneurship, innovation, and transformation. *Organization & Environment*, 29(3), 264–289

Schiavo-Campo, S., & Cordes, J. J. (2017). *The role of storage in energy supply*. Routledge.

Stern, P. C. (2000). New environmental theories: toward a coherent theory of environmentally significant behavior. *Journal of Social Issues*, 56(3), 407–424.

Swan, M. (2015). *Blockchain: Blueprint for a new economy*. O'Reilly Media.

Szulanski, G. (1996). Exploring internal stickiness: Impediments to the transfer of best practice within the firm. *Strategic Management Journal*, 17(S2), 27–43.

Tietze, F., Alcántara, M. G., Eichhorn, M., & Harms, L. (2020). *Electric mobility: Innovation, sustainability, and governance*. Springer.

UN Environment. (2018). *UN environment annual report 2018*. United Nations Environment Programme.

UNESCO. (2005). *Education for sustainable development: A roadmap*. United Nations Educational, Scientific and Cultural Organization.

Vanclay, F., Esteves, A. M., Aucamp, I., Franks, D. M., & Griffiths, M. (2015). *Social impact assessment: Guidance for assessing and managing the social impacts of projects*. International Association for Impact Assessment.

Wasko, M. M., & Faraj, S. (2005). Why should I share? Examining social capital and knowledge contribution in electronic networks of practice. *MIS Quarterly*, 29(1), 35–57.

Wenger, E. (1998). Communities of practice: Learning, meaning, and identity. Cambridge University Press.

Wenger, E., McDermott, R., & Snyder, W. M. (2002). Cultivating communities of practice: A guide to managing knowledge. Harvard Business Press.

Wolfert, S., Ge, L., Verdouw, C., & Bogaardt, M. J. (2017). Big data in smart farming—A review. *Agricultural Systems*, 153, 69–80.

Xiong, X., Zheng, K., Lei, L., & Hou, L. (2020). Resource allocation based on deep reinforcement learning in IoT edge computing. *IEEE Journal on Selected Areas in Communications*, 38(6), 1133–1146.

Yang, M., Ren, Y., & Adomavicius, G. (2019). Understanding user-generated content and customer engagement on Facebook business pages. *Information Systems Research*, 30(3), 839–855.

Zahra, S. A., & George, G. (2002). The net-enabled business innovation cycle and the evolution of dynamic capabilities. *Information Systems Research*, 13(2), 147–150.

9 New Economic Models for Sustainability-Oriented Knowledge Transfer?

A Contrasting Case Study on Two Knowledge Transfer Projects

Daniel Feser

Introduction

This chapter contributes to the discussion of the European environmental "conscience" by exploring academic knowledge transfer. Specifically, in policy programmes like the European Green deal (European Commission, 2019), higher education institutions (HEIs) have been discussed as an important actor in the European knowledge economy. Research also provides insights that HEIs influence sustainability on a European level (Disterheft et al., 2012; Lozano et al., 2022). This chapter examines two dimensions of European environmental "conscience" being connected to HEIs: First, the "innovative nature of solutions, or a commitment to progress" (Hoerber and Weber, 2021, p. 2) and second, in relation with "[g]rassroots support and demand for effective solutions" (Hoerber and Weber, 2021, p. 2).

Knowledge transfer traditionally targets companies as central actors commercialising new knowledge into business models, products and services (Carlsson and Fridh, 2002; Vinig and Lips, 2015). Undoubtedly, patenting has been in the core of the discussion of knowledge and technology transfer (Asna Ashari et al., 2023; Hall, 2014; Hall and Helmers, 2010; Owen-Smith and Powell, 2001). Often literature focuses on the connection of knowledge transfer to economic development and growth (Acs et al., 2012; Audretsch, 1995).

The contribution of knowledge transfer towards sustainability has just recently gained more attention (Hall and Helmers, 2010). One main reason is the commitment to accelerating sustainability of a large share of the UN member states in international treaties like the Paris agreement and the Sustainable Development Goals. Fighting the persisting sustainability crisis and overcoming the "wicked" problems have led to the need to reformulate and reframe discussion in innovation and entrepreneurial ecosystem research (Bischoff and Volkmann, 2018; Levin et al., 2012). This resulted in a change of focus from R&D and systems innovation approaches to transformative frameworks, the so-called innovation policy 3.0 (Schot and Steinmueller, 2018). Concepts like the circular economy and green tech innovation have been prominently promoted (e.g.Smith, 2007; Stahel, 2016) and inherently demonstrated the need to foster sustainability-oriented innovation (SOI). Consequently, directionality

DOI: 10.4324/9781032656359-13

of innovation and knowledge exchange processes require a review of the knowledge economy to fully understand the role of SOI in accelerating sustainability (Edler and Boon, 2018; Motta, 2018).

Knowledge transfer as source of acceleration to the diffusion of SOI has become again a relevant discussion topic. Especially, innovation intermediaries play an important role by boundary spanning and technology and knowledge broking (Howells, 2006). The intermediation of knowledge between universities and companies has been described as particularly difficult due to different knowledge production modes, use of knowledge bases and innovative capacities (Alunurm et al., 2020; Bruce S. Tether and Abdelouahid Tajar, 2008).

The SOI literature conceptualises university as a change agent for supporting and fostering sustainable entrepreneurial activities (Stephens et al., 2008; Wakkee et al., 2019). With the focus of this paper on the intermediation between university and companies, the notion of a "black box" (Ferreira and Carayannis, 2019, p. 353) has been used to clarify how much needs to be explored to fully understand benefits and obstacles of knowledge transfer. Volkmann et al. (2021) proposed with the sustainable entrepreneurial ecosystems a framework to conceptualise directed entrepreneurial activities for SOI using an explicit ecosystem perspective and incorporating the relevance of knowledge spillovers.

This chapter analyses the function of knowledge transfer in sustainable entrepreneurial ecosystems contributing to the European environmental "conscience". However, we still know very little about how knowledge transfer targeting sustainable development differs from traditional approaches. Following the demand of Wagner et al. (2019, p. 15) "research more on the process and effect dimensions", this paper focuses on differences in processes between undirected and directed knowledge transfer for SOI. While I assume that both approaches can lead to SOI, knowledge about the specific activities is scarce. The upcoming but still small research strand has focused on knowledge spillover (Wagner et al., 2019), the interaction between entrepreneurial ecosystem (Franco-Leal et al., 2020) and the role of social capital (Theodoraki et al., 2018). My research adds two dimensions to the currently discussed topics. First, I will concentrate on the difference between "traditional" undirected and directed knowledge transfer approaches. Second, the underlying institutional conditions necessary to successfully embed academic knowledge in entrepreneurial systems and accelerate commercialisation will be a central contribution of this paper.

This paper aims to answer the research question, how do knowledge transfer offices (KTOs) shape knowledge exchange for SOI in sustainable entrepreneurial ecosystems? This paper applies an inductive and exploratory approach and analyses a set of 32 exploratory interviews with stakeholder at the German Applied University of Darmstadt (AUD) and University of Goettingen (UG), which are involved in knowledge transfer in their respective entrepreneurial system. I use a contrasting case study approach. While the knowledge transfer of AUD has an explicit focus on SOI, the UG has applied a "traditional" approach for their knowledge transfer activities aiming to foster the general regional innovativeness.

The remainder is structured as follows. Section 2 explains the literature background including the literature on SOI, university knowledge transfer in entrepreneurial ecosystems and institutional analysis. Section 3 describes the methodology and data. Furthermore, Sections 4 and 5 report the results of the two comparative case studies and Section 6 discusses the relevance of the results regarding their relevance to specific literature strands. Section 6 concludes with a brief summary and limitation.

Literature Background

This section offers a brief overview of the literature relevant to the research question relating the European conscience dimensions of innovative progress and grassroots in transfer (Hoerber and Weber, 2021). In particular, I review theoretical and empirical insights on the emergence and requirements of SOI, the role of universities in entrepreneurial ecosystems and institutional theory.

Sustainability-oriented Innovation and Knowledge Transfer

Klewitz and Hansen (2014) define in their literature review SOI "as a 'direction', which to follow requires the deliberate management of economic, social, and ecological aspects [..] so that they become integrated into the design of new products, processes, and organizational structures" (Klewitz and Hansen, 2014, p. 57).[1] Historically, innovation research has employed frameworks which support the innovation process as decentral and undirected. In this context, negative effects towards undirected or falsely directed innovation has been largely neglected (Røpke, 2012). As sustainability has become a more relevant dimension of innovation, research has started a fruitful discussion to integrate this perspective in its research framework. Literature has discussed factors influencing and supporting SOI (Hojnik and Ruzzier, 2016), with focus on SMEs (Hansen et al., 2002; Klewitz and Hansen, 2014), green industries (Grillitsch and Hansen, 2019), collaboration with administrative actors (Ball et al., 2018) and different sub-groups of SOI-Innovation, like sustainable product, process, organisational and business model innovation (Adams et al., 2016; Geissdoerfer et al., 2018; Klewitz and Hansen, 2014). A core result of the literature contains the inherent complexity of SOI and the need to cooperate with external partner to improve the impact on sustainable development. In particular, SMEs and entrepreneurial activities have been addressed to understand the underlying systemic mechanisms.

Knowledge Transfer and Universities in Entrepreneurial Ecosystems

The concept of entrepreneurial ecosystem is based on the knowledge spillover theory of entrepreneurship, which discusses the role of the knowledge economy for entrepreneurship (Ács, 2009; Audretsch and Keilbach, 2007). Entrepreneurial ecosystems have been defined as "the combination of social, political, economic, and cultural elements within a region that support the development and growth of innovation-based ventures" (Civera et al., 2019, p. 381). Universities have been attributed a central role in entrepreneurial ecosystems (Audretsch and Link, 2017). Academic knowledge is not only seen as a source for the ecosystem but has also been increasingly discussed as important due to reciprocal linkages and as a provider of entrepreneurial activities in recent years (Cerver Romero et al., 2020; Lockett et al., 2003). Studies show that universities foster international knowledge spillover via academic spillover (Civera et al., 2019), contribute with unintended knowledge spillover to the foundation of high tech firms (Acosta Seró et al., 2011) and are target of regional policy to foster regional development (Benneworth and Charles, 2005). A central activity is the intermediation between universities and other actor in the entrepreneurial ecosystems (Hayter, 2016). Universities are particulary important for boundary spanning and network creation between academic institution and entrepreneurial ecosystem (Comacchio et al., 2012; Hayter, 2016).

Specifically, the KTOs as organisational units at universities have received much attention supporting the creation, capturing and diffusion of knowledge (Alexander and Martin, 2013; Cunningham and O'Reilly, 2018; Siegel et al., 2003). While a common definition has yet to be established and depends on the focus of specific HEIs, KTOs are assumed to be central organisational units responsible for fostering, conducting, mediating, and facilitating all forms of knowledge (Zhou and Tang, 2020). KTOs have become a driving force to support entrepreneurial activities, like networking between entrepreneurs, consulting of founders and foster cooperation for innovation, such as helping with IPR licensing, e.g. commercialising patents (Drivas et al., 2016; Owen-Smith and Powell, 2001). In entrepreneurial ecosystems, KTOs might potentially serve as change agents, influencing dynamics on a systemic level. Nevertheless, empirical evidence and insights from the micro level are still missing, even though the importance of understanding university transfer strategies has been emphasised (Caldera and Debande, 2010; Feldman et al., 2002; Horner et al., 2019).

Concerning the research question, the literature has expanded the concept of entrepreneurial ecosystems with an emphasis on the sustainability and SOI (Tiba et al., 2020). Still in a premature state, my contribution can help to identify and understand the role of universities better. Little is known about how the process of directed knowledge flows between universities and entrepreneurial actors works and how contextual conditions shape them. The next section uses insights from economic theory to conceptualise the research problem.

Knowledge Dissemination and Club Goods

To begin with, the famous knowledge paradox (Arrow, 1962), concerning difficulties successfully commercialising knowledge between different actors reflects a major problem because knowledge incorporates characteristics of public goods. Being not excludable and non-rival, the dissemination of knowledge is affected by lack of incentives and free rider problems. Moreover, this becomes even more challenging for knowledge transfer when dealing with environmental problems referring to the concept of the "tragedy of the commons" (Hardin, 1968). Overuse and inefficiency in common-pool resources requires solutions beside simplistic market solutions (Ostrom, 1999). Particularly in the context of directional knowledge transfer, the connection between universities and businesses requires further revision. This is especially true when it comes to the "double externality" (Hall and Helmers, 2010, p. 2) which adds to the knowledge transfer sustainability challenge, such as solving environmental pollution problems.

To overcome problems with public goods, limiting access and turn the good into an excludable good is widely used. Club goods are characterised by goods that are excludable but not rival. On the basis of Buchanan's (1965) seminal work, a distinct discussion on artificially limiting access to public goods can help to overcome market failure and support the efficient provision of club goods by markets. For the knowledge transfer, the discussion has begun for Intellectual Property Rights, like patents, which incentivise knowledge production and the diffusion of innovation (Schwartz, 2017). Nevertheless, research is just in the premature phase, when it comes to an economic analysis concerning the transition of public to club goods. A deeper understanding of the knowledge economy can be expected and helps to understand how club good mechanism can help to foster the diffusion of SOI.

Institutional Analysis for Analysing Sustainable Entrepreneurial Ecosystems

Based on Bizer and Führ (2015), I will explore the incentives, heuristics and embeddedness of universities in knowledge transfer for SOI. Different outputs and outcomes of entrepreneurial ecosystems are the main motive for this strand of literature (Bischoff and Volkmann, 2018). Institutional analysis can help to understand better the function of systems. Mismatches and misconception of institutional frameworks have been identified for the emergence of SOI (Smink et al., 2015). Literature focusing on dynamic processes for SOI emphasise on the misconception of institutional frameworks as assuming them to static and homogeneous (Fuenfschilling and Truffer, 2014, 2016). Institutional change has been traditionally incorporated in the analysis of universities and their role in innovation systems (Benneworth et al., 2017). Understanding the formal and informal set of rules in knowledge transfer can further help uncover the functions and mechanisms within entrepreneurial ecosystems.

Method and Data

This study applies a case study approach since currently the literature is in the exploratory state and a qualitative approach can help to discuss "how"-related research questions (Yin, 2003) to learn about processes of knowledge transfer between businesses and universities as part of European environmental "conscience" (Hoerber and Weber, 2021). Case study research has been commonly used to analyse the HEIs and knowledge transfer (Benneworth et al., 2017; Tiemann et al., 2018). This inductive study aims to gain theoretical insights (Eisenhardt, 1989) on knowledge transfer for SOI with the help of contrasting the two different universities in Germany. Specifically, a contrasting case study approach was used to create a rich data set and disclose new knowledge (e.g. Geels and Kemp, 2007; Swamidass, 2013). For this purpose two different innovation policy projects at universities were selected which aim to contribute regional sustainable development.

As part of the project IreWiNE (Indicators about regional knowledge transfer for sustainable development), the UG and the AUD were chosen since both universities pursued contrasting knowledge transfer projects goals and followed different ways to achieve their project goals.[2] The project of the UG, called SNIC (Southern Lower Saxony Innovation Campus), was built to support knowledge exchange in the region. The UG has a well-perceived international reputation. For example, more than 40 researchers linked to the UG have won the Nobel Prize. However, the regional entrepreneurial ecosystem is only loosely related to the university. Academic knowledge transfer is not perceived as influential, according to interviewees, since a large share of the research output has merely a focus on basic research. In recent years, the KTO focuses on connecting relevant regional players. The SNIC solely focused on innovation transfer with emphasis on regional development and improving knowledge transfer in the entrepreneurial ecosystem.

The project s:ne (Systemic Innovation for sustainable Development) at the AUD focuses on changing knowledge transfer at the university from a general innovation perspective to a SOI perspective. Traditionally the AUD is specialised on educating students with a majority of technical and engineering students. Furthermore, applied knowledge transfer is focus of the transfer strategy of the AUD. Knowledge transfer as understood in s:ne explicitly should contribute to SOI and direct the activities of the university towards more sustainability. Furthermore, the university as opposite to the UG approach is embedded in a metropolitan innovation system, which includes large companies, leading research institutions on environmental and sustainability topics and a lively entrepreneurship ecosystem.

An explorative approach using semi-structured interviews was selected. The IreWiNE research project team developed a semi-structured guideline within the project team. We initiated the process with desk research and conducted preliminary testing of the questionnaire. Throughout the interview phase, researchers engaged in feedback rounds to ensure that the questions adequately addressed the relevant content. In conducting qualitative research, the questionnaire predominantly comprised open and semi-open questions to ensure unbiased responses and to glean insights from respondents (See Table 9.1).

Table 9.1 Topics of interview guide

Section 0: Background of the interviewee
Section 1: Knowledge transfer structures and characteristics of key stakeholders
Section 2: Innovation processes
Section 3: Evaluation and assessment of results
Section 4: The regional innovation system
Section 5: Sustainable development

The 32 interviews (15 interviews with focus on the UG and 17 interviews on the AUD) were conducted with regional innovation experts which were directly, like in applied projects with regional stakeholders and indirectly, like organising networking activities and practitioner conferences involved in the regional knowledge transfer and had expertise concerning SOI (see Table 9.2). The selection of the interviewees followed the procedure of the theoretical sampling conducting interviews until saturation of the interviews (Glaser and Strauss, 1967). The project team members started with the directly involved responsible persons and used than snowballing to collect more insights, also from external actors being active in the region. Specifically, internal and external persons from research, companies, administration and civil society were interviewed in both regions to receive a full picture on SOI-relevant knowledge. These interviews were conducted by the IreWiNE-research team between February and July 2020 via video call due to social distance restrictions during the Covid19-Pandemic. The interviews lasted between 31 Min. and 1:51h. Subsequently, the interviews were recorded, transcribed, the codes discussed in the research team and preliminary results in stakeholder webinars presented and feedback involved in the analysis.

The analysis follows the procedure of content analysis (Mayring, 2015). Content analysis aims to distil the essence of the text material and cross-check the interviewees' statements. The analysis followed a two-stage procedure. First, the interviews were examined within each region. Second, a cross-case analysis was conducted to obtain comparable results. The analysis involved a mixture of inductive coding for new insights and deductive coding with references from the literature mainly stemming from knowledge transfer and club good theory. Furthermore, additional publications, related webpages, and other materials on the two regions and the knowledge transfer activities were used to ensure the reliability of the interviews and the cases. To ensure the results were comprehensive and reflected diverse perspectives, within the project facilitated discussions in stakeholder workshops and a conference took place. Participants included representatives from both region regions.

Results: Knowledge Transfer for SOI As Club Good

In this section, I analyse the prerequisites of knowledge transfer for sustainable development, focusing on the characteristics of the economic form of the

Table 9.2 Interviews conducted

No.	Region	Sector	Role	Perspective on university	Duration (min)
1	Darmstadt	Academia	Professor (s:ne team member)	Internal	112
2	Darmstadt	Academia	Research associate (s:ne team member)	Internal	40
3	Darmstadt	Civil society	Research associate of a foundation (s:ne team member)	External	54
4	Darmstadt	Industry	Representative of chamber of commerce	External	61
5	Darmstadt	Academia	Research associate of a research institute (s:ne team member)	External	66
6	Darmstadt	Academia	Research associate of a research institute	Internal	36
7	Darmstadt	Academia	Research associate (s:ne team member)	Internal	65
8	Darmstadt	Academia	Senior researcher of a research institute (s:ne team member)	Internal	91
9	Darmstadt	Academia	Senior researcher (s:ne team member)	External	90
10	Darmstadt	Industry	Representative of chamber of commerce	External	91
11	Darmstadt	Public administration	Innovation support and technology transfer manager	External	34
12	Darmstadt	Industry	Representative of business association	External	40
13	Darmstadt	Academia	Senior researcher (s:ne team member)	Internal	58
14	Darmstadt	Industry	Sustainability consultant (s:ne team member)	External	57
15	Darmstadt	Academia	Representative of university sustainability office	External	45
16	Darmstadt	Academia	Representative of university presidential board	Internal	59
17	Darmstadt	Public administration	Representative of university transfer office	Internal	42
18	Göttingen	Academia	Professor (SNIC team member)	Internal	87
19	Göttingen	Academia	Project manager (SNIC team member)	Internal	92
20	Göttingen	Public administration	Representative of SNIC Office (SNIC team member)	External	72
21	Göttingen	Public administration	Innovation support and technology transfer manager (SNIC team member)	External	62

22	Göttingen	Civil society	Representative of a foundation	External	59
23	Göttingen	Academia	Professor (SNIC team member)	External	75
24	Göttingen	Public administration	Business developer (SNIC team member)	External	64
25	Göttingen	Public administration	Business developer (SNIC team member)	External	48
26	Göttingen	Academia	Research associate (SNIC team member)	External	53
27	Göttingen	Academia	Innovation scout (SNIC team member)	External	50
28	Göttingen	Industry	Representative of chamber of crafts	External	71
29	Göttingen	Public administration	Business developer (SNIC team member)	External	50
30	Göttingen	Industry	Representative of chamber of commerce	External	78
31	Göttingen	Academia	Innovation scout (SNIC team member)	Internal	55
32	Göttingen	Public administration	Business developer (SNIC team member)	External	52

knowledge being transferred between universities and companies. The solutions discussed in the interviews encompassed the advancement of new technologies, such as enhancing sustainable production processes, as well as methodological innovations like participatory stakeholder strategy development for urban development and within the leather industry.

Knowledge Transfer as Club Good

Requirements from public policy to contribute to knowledge generation, in addition to education and research, have established the need for the so-called "third mission" of universities, which involves actively engaging in knowledge transfer. The interviewees in both regions stated problems for fostering knowledge transfer and perceived particular challenges when it comes to SOI. AUD and OG receive public funding to strategically reformulate goals, implement new processes, and foster regional knowledge transfer activities. In both cases, the universities' interactions with businesses were key to overcoming low demand for academic knowledge and tackling the low interest of actors collaborating with academic entities due to high transaction costs and uncertainty about the novelty and usability of the knowledge offered. This was partially due to the characteristics of public goods. While the AUD reacted by innovating their strategy (as discussed in the subsequent section), the UG intensified activities to expand the network within the entrepreneurial ecosystem and increase the number of prospective cooperation partners for innovation projects.

Sustainability-oriented Innovation as Excludable Goods

The directionality of knowledge transfer only plays an explicit role in the innovation strategy of knowledge transfer at the AUD. AUD has decided to implement a new strategic approach to prioritise knowledge transfer for sustainability. The project "s:ne" was set up to support the internal knowledge transfer units and stakeholders at the AUD to improve the impact of innovation towards sustainability. Sustainability is operationalised as impacting the Sustainable Development Goals (SDGs). In contrast, at the SNIC, while highly recognised research on sustainability-related topics is conducted, SOI is not explicitly prioritised. However, SOI projects still play a prominent role in knowledge transfer projects.

The strategy towards a more sustainable impact has an inherent exclusion mechanism to collaborate with SOI change agents. Specific actors, which have traditionally played a more important role in knowledge transfer, such as craft and industrial chambers and professional organisations, have explicitly not been selected due to the perception that the focus of SOI would not be understood as an important priority. Instead, collaboration partners were selected according to the organisations' explicit focus and their experience guiding, diffusing and creating SOI. Furthermore, best practice knowledge transfer projects have been developed with the precondition that actors require time and knowledge to participate at the knowledge intermediation processes. These processes aim to identify SOI with the potential to induce systemic change. They include strategy processes involving entrepreneurs for more than one year. For example, in the case of leather companies and sustainable mobility, these processes started with goal-setting and strategy development, including creativity exercises for collaboratively developing SOI. While the exclusion mechanism has not been officially implemented (due to ethical and legal issues), the analysis of the project indicates a clear intention to foster intensive collaboration with potentially successful participants. This careful designed selection mode was argued to be necessary to cooperate with motivated, knowledgeable actors, which also have enough resources to induce systemic changes via SOI and provide the targeted institutional framework to diffuse in entrepreneurial ecosystems.

The UG on the opposite hand pursues an explicit open participation approach. Specifically, the regional municipalities with public economic development units were selected and introduced to the network to access a geographical broader scope. Furthermore, other closely related universities were integrated in the knowledge transfer concept to broaden the basis and open the networks for many different participating companies.

Results: Influencing Factors

In this section, I report on the factors influencing knowledge transfer for sustainability, which influence the network capacity building for SOI and support changes towards sustainable development in knowledge transfer organisations. The interviewees highlighted especially the internal-university networks, regional context, and dynamics as relevant for the capacity to orient successfully towards SOI.

Internal-university Networks

In Darmstadt, professors play a central role in knowledge transfer. Professors often have experience in industry jobs and use network resources to initiate network projects since often in university of applied sciences are recruited directly from the industry. The explicit applied focus of the AUD promotes these collaborations. In recent years, a group of researchers, students, and administrators focusing on sustainability-related topics has emerged. They are organised in a group called i:ne (English translation: Initiative: Sustainable Development at the AUD). The interviewees describe a well-connected informal network at the AUD that has the capacity to gather different competences and skills to enable SOI projects.

Consequently, an interdisciplinary research and knowledge transfer centre on sustainability research has been founded to collect relevant activities in Darmstadt. This structure has had a positive impact on fostering sustainability goal-setting inside the university and implementing a knowledge transfer strategy, which aims to contribute with SOI to the SDGs. It also entails an explicit idea that sustainable knowledge transfer needs to take into account not only the traditional metrics but also an orientation on sustainability-related impact.

Regional Context

Comparing both case regions, different regional entrepreneurial ecosystems shape the outcomes of knowledge transfer projects. The exclusive orientation towards SOI has been identified as challenging for UG. In particular, the low density of actors with specialised knowledge and few knowledge networks with a sustainability focus have led to a low visibility of sustainability in the Goettingen region. In contrast, Darmstadt has an ecosystem with various well-known actors collaborating in SOI. For example, the Passive House Institute in Darmstadt has established a world-wide low energy standard for buildings and has close ties to the AUD. Specifically, when connecting with local stakeholder groups and grassroots movements like the student sustainability initiative at the university, known as i:ne, it becomes crucial to consider the regional context. According to interviewees at the AUD, searching and developing new interaction methods require an understanding of varying success rates and responsive learning techniques that align with the regional context.

Dynamics

The dynamics of network development have been discussed by the interviewees as crucial. Both projects have built on previous initiatives and established connections with other relevant actors within the entrepreneurial ecosystems. The involved actors in the projects emphasise the importance of knowledge networks and their connections to other actors within the region. In the case of AUD, internal sustainability-focused networks have been developed for more than thirty years, including various projects concerning SOI and the design of entrepreneurial transformation scenarios. Furthermore, the preparation of the projects, with a focus on university-wide implementation of knowledge transfer towards sustainability, has significantly benefited from these internal university networks.

Discussion

To my knowledge, the role of clubs for knowledge transfer has not yet been discussed concerning their role in supporting the knowledge exchange for SOI. At the international level, clubs have been recognised as means to accelerate international cooperation for climate change policy (Hovi et al., 2016, 2019; Platje and Kampen, 2016; Sprinz et al., 2018). While this literature does not specifically discuss the role of knowledge exchange, the same mechanism of exclusively including advanced cooperation of front-runner and change actors creates clubs to accelerate sustainable transitions. My contribution can add to the discussion on sustainable club settings with a specific regional and micro-perspective. The discussion on club goods and the institutional transformation from public goods to club goods has been discussed in economics at a rather theoretical level. The organisation of knowledge transfer as a club good can help to understand how institutional frameworks need to be designed to incentivise actors to contribute to SOI and benefit from influencing sustainability.

Furthermore, the results of this study contribute to the contextualisation of sustainable entrepreneurial ecosystems (Bischoff and Volkmann, 2018; Volkmann et al., 2021). It is specifically relevant to understand the role of universities in supporting regional sustainable development. The findings suggest that for understanding competitive advantage, contextual conditions, such as the dynamics of knowledge networks, density of regional organisations with a sustainability focus, and internal networks, can be explanatory factors that influence different outcomes in entrepreneurial ecosystems. Furthermore, institutional analysis is needed to conduct more in-depth research.

Conclusion

This chapter presents a contrasting case study on AUD and UG with qualitative expert interviews on the role of knowledge transfer for European consciences, focusing on innovative solutions and HEIs' bottom-up support. The literature has discussed underlying factors for successful SOI, barriers to knowledge transfer, and the role of entrepreneurial ecosystems. However, insights are scarce on universities' strategies to accelerate the diffusion of SOI and the directionality induced changes for the universities. This study investigates the research question of how knowledge transfer is shaped to support and accelerate SOI in a university setting. First, I find a strategy to redefine knowledge transfer for SOI as a club good. The added directionality has been used to establish novel club good structures and to incentivise actors to contribute to sustainable transitions by making collaboration more attractive for active participants. This approach has been chosen since the AUD aims for SOI on a systemic level and targets explicitly for disruptive innovation while UG attempts to increase the number of incremental innovation projects influencing regional development. Second, internal networks as well as regional connections of universities shape the institutional framework of knowledge transfer. In the cases of AUD and UG, different contexts exist. Consequently, a one-size-fits-all approach seems not to be a successful strategic approach. Third, for building a club for sustainable dynamic development of network structure, it has been a prerequisite to follow an approach that allows to barrier the access to the university's services.

This study adds to conceptualising the role of institutions in sustainable knowledge transfer. Given the inductive exploratory approach, more case studies on different universities, regions, and entrepreneurial ecosystems will be helpful to enrich the knowledge about the transition of knowledge transfer towards SOI, supporting sustainable entrepreneurial ecosystems. Nevertheless, taking into account my results, respectively the role of club good characteristics and institutional framework, directionality requires more reflection in innovation studies since processes, as demonstrated, change when introducing explicit normative focus on sustainability in university-business interaction. Furthermore, regarding environmental awareness, universities can serve as intermediaries in fostering environmental conscience. As demonstrated in this chapter, universities function as regional hubs connecting grassroots groups with scientific stakeholders and can also drive innovation development. Exploring the interplay between energy policies, environmental initiatives, and their impact on European institutions such as the European Commission presents promising avenues for future research on the role of universities in the European conscience.

Notes

1 There are a large number of different definition, which rely on different conceptualization, for a comprehensive critique see Adams et al. (2016); Klewitz and Hansen (2014) SOI was chosen since it connects the existing innovation framework with policy goals like the SDGs.
2 For further insights, see related publications using this data set: Bäumle et al. (2023b); Bäumle et al. (2023a); Friedrich and Feser (2023).

Bibliography

Acosta Seró, M., Coronado Guerrero, D., Flores, E., 2011. University spillovers and new business location in high-technology sectors: Spanish evidence. *Small Business Economics* 36 (3), 365–376.

Ács, Z.J., 2009. *The knowledge spillover theory of entrepreneurship*, Springer. https://doi.org/10.1007/s11187-010-9307-2

Acs, Z.J., Audretsch, D.B., Braunerhjelm, P., Carlsson, B., 2012. Growth and entrepreneurship. *Small Business Economics* 39 (2), 289–300.

Adams, R., Jeanrenaud, S., Bessant, J., Denyer, D., Overy, P., 2016. Sustainability-oriented innovation: A systematic review. *International Journal of Management Reviews* 18 (2), 180–205. doi:10.1111/ijmr.12068

Alexander, A.T., Martin, D.P., 2013. Intermediaries for open innovation: A competence-based comparison of knowledge transfer offices practices. *Technological Forecasting and Social Change* 80 (1), 38–49. doi:10.1016/j.techfore.2012.07.013

Alunurm, R., Rõigas, K., Varblane, U., 2020. The relative significance of higher education–industry cooperation barriers for different firms. *Industry and Higher Education*. doi:10.1177/0950422220909737

Arrow, K., 1962. Economic welfare and the allocation of resources for invention. In Nelson, R.R. (Ed), *The rate and direction of inventive activity: Economic and social factors*. Princeton University Press, pp. 609–626.

Asna Ashari, P., Blind, K., Koch, C., 2023. Knowledge and technology transfer via publications, patents, standards: Exploring the hydrogen technological innovation system. *Technological Forecasting and Social Change* 187, 122201. doi:10.1016/j.techfore.2022.122201

Audretsch, D., Link, A., 2017. *Universities and the entrepreneurial ecosystem*. Edward Elgar Publishing.

Audretsch, D.B., 1995. *Innovation and industry evolution*. MIT Press.

Audretsch, D.B., Keilbach, M.C., 2007. The theory of knowledge spillover entrepreneurship. *Journal of Management Studies* 44 (7), 1242–1254.

Ball, C., Burt, G., de Vries, F., MacEachern, E., 2018. How environmental protection agencies can promote eco-innovation: The prospect of voluntary reciprocal legitimacy. *Technological Forecasting and Social Change* 129, 242–253. doi:10.1016/j.techfore.2017.11.004

Bäumle, P., Hirschmann, D., Feser, D., 2023a. The contribution of knowledge intermediation to sustainability transitions and digitalization: Qualitative insights into four German regions. *Technology in Society* 73, 102252. doi:10.1016/j.techsoc.2023.102252

Bäumle, P., Hirschmann, D., Feser, D., Winkler-Portmann, S.J., Bizer, K., 2023b. *Digitalisierung und regionaler Wissenstransfer: Interdependenzen und Herausforderungen*. Standort. doi:10.1007/s00548-023-00844-3

Benneworth, P., Charles, D., 2005. University spin-off policies and economic development in Less successful regions: Learning from two decades of policy practice. *European Planning Studies* 13 (4), 537–557. doi:10.1080/09654310500107175

Benneworth, P., Pinheiro, R., Karlsen, J., 2017. Strategic agency and institutional change: Investigating the role of universities in regional innovation systems (RISs). *Regional Studies* 51 (2), 235–248. doi:10.1080/00343404.2016.1215599

Bischoff, K., Volkmann, C.K., 2018. Stakeholder support for sustainable entrepreneurship: A framework of sustainable entrepreneurial ecosystems. *International Journal of Entrepreneurial Venturing* 10 (2), 172. doi:10.1504/IJEV.2018.092714

Bizer, K., Führ, M., 2015. Compact guidelines: Practical procedure in interdisciplinary institutional analysis. *sofia-Diskussionsbeiträge* 15 (4), 1–22. Darmstadt - https://doi.org/10.46850/sofia.9783941627451s

Buchanan, J.M., 1965. An economic theory of clubs. *Economica* 32 (125), 1. doi:10.2307/2552442

Caldera, A., Debande, O., 2010. Performance of Spanish universities in technology transfer: An empirical analysis. *Research Policy* 39 (9), 1160–1173. doi:10.1016/j.respol.2010.05.016

Carlsson, B., Fridh, A.-C., 2002. Technology transfer in United States universities. *Journal of Evolutionary Economics* 12 (1-2), 199–232. doi:10.1007/s00191-002-0105-0

Cerver Romero, E., Ferreira, J.J.M., Fernandes, C.I., 2020. The multiple faces of the entrepreneurial university: A review of the prevailing theoretical approaches. *The Journal of Technology Transfer*. doi:10.1007/s10961-020-09815-4

Civera, A., Meoli, M., Vismara, S., 2019. Do academic spinoffs internationalize? *The Journal of Technology Transfer* 44 (2), 381–403. doi:10.1007/s10961-018-9683-3

Comacchio, A., Bonesso, S., Pizzi, C., 2012. Boundary spanning between industry and university: The role of technology transfer centres. *Journal of Technology Transfer* 37 (6), 943–966. doi:10.1007/s10961-011-9227-6

Cunningham, J.A., O'Reilly, P., 2018. Macro, meso and micro perspectives of technology transfer. *The Journal of Technology Transfer* 43 (3), 545–557. doi:10.1007/s10961-018-9658-4

Disterheft, A., Da Ferreira Silva Caeiro, S.S., Ramos, M.R., de Miranda Azeiteiro, U.M., 2012. Environmental Management Systems (EMS) implementation processes and practices in European higher education institutions – Top-down versus participatory approaches. *Journal of Cleaner Production* 31, 80–90. doi:10.1016/j.jclepro.2012.02.034

Drivas, K., Economidou, C., Karamanis, D., Zank, A., 2016. Academic patents and technology transfer. *Journal of Engineering and Technology Management* 40, 45–63. doi:10.1016/j.jengtecman.2016.04.001.

Edler, J., Boon, W.P., 2018. 'The next generation of innovation policy: Directionality and the role of demand-oriented instruments'—Introduction to the special section. *Science and Public Policy* 45 (4), 433–434. doi:10.1093/scipol/scy026.

Eisenhardt, K.M., 1989. Building theories from case study research. *The Academy of Management Review* 14 (4), 532–550.

European Commission, 2019. The European Green Deal. COM(2019) 640 final, Brussels (downloaded on 28 August 2020 from https://eur-lex.europa.eu/legal-content/EN/TXT/HTML/?uri=CELEX:52019DC0640&from=EN)

Feldman, M., Feller, I., Bercovitz, J., Burton, R., 2002. Equity and the technology transfer strategies of American Research Universities. *Management Science* 48 (1), 105–121. doi:10.1287/mnsc.48.1.105.14276

Ferreira, J.J.M., Carayannis, E.G., 2019. University-industry knowledge transfer—Unpacking the "black box": An introduction. *Knowledge Management Research and Practice* 17 (4), 353–357. doi:10.1080/14778238.2019.1666514.

Franco-Leal, N., Camelo-Ordaz, C., Dianez-Gonzalez, J.P., Sousa-Ginel, E., 2020. The role of social and institutional contexts in social innovations of Spanish academic spinoffs. *Sustainability* 12 (3), 906. doi:10.3390/su12030906

Friedrich, C., Feser, D., 2023. Combining knowledge bases for small wins in peripheral regions. An analysis of the role of innovation intermediaries in sustainability transitions. *Review of Regional Research*. doi:10.1007/s10037-023-00192-7

Fuenfschilling, L., Truffer, B., 2014. The structuration of socio-technical regimes—Conceptual foundations from institutional theory. *Research Policy* 43 (4), 772–791. doi:10.1016/j.respol.2013.10.010

Fuenfschilling, L., Truffer, B., 2016. The interplay of institutions, actors and technologies in socio-technical systems—An analysis of transformations in the Australian urban water sector. *Technological Forecasting and Social Change* 103, 298–312. doi:10.1016/j.techfore.2015.11.023

Geels, F.W., Kemp, R., 2007. Dynamics in socio-technical systems: Typology of change processes and contrasting case studies. *Technology in Society* 29 (4), 441–455. doi:10.1016/j.techsoc.2007.08.009

Geissdoerfer, M., Vladimirova, D., Evans, S., 2018. Sustainable business model innovation: A review. *Journal of Cleaner Production* 198, 401–416. doi:10.1016/j.jclepro.2018.06.240

Glaser, B.G., Strauss, A.L., 1967. *The discovery of grounded theory: Strategies for qualitative research*. Aldine

Grillitsch, M., Hansen, T., 2019. Green industry development in different types of regions. *European Planning Studies* 27 (11), 2163–2183. doi:10.1080/09654313.2019.1 648385

Hall, B.H., 2014. Does patent protection help or hinder technology transfer? In Ahn, S., Hall, B.H., Lee, K. (Eds.), *Intellectual property for economic development*. Edward Elgar Publishing

Hall, B.H., Helmers, C., 2010. The role of patent protection in (clean/green) technology transfer. NBER Working Paper (16323).

Hansen, O.E., Søndergård, B.B., Meredith, S., 2002. Environmental innovations in small and medium sized enterprises. *Technology Analysis & Strategic Management* 14 (1), 37–56. doi:10.1080/09537320220125874

Hardin, G., 1968. The tragedy of the commons. *Science (New York, N.Y.)* 162 (3859), 1243–1248.

Hayter, C.S., 2016. A trajectory of early-stage spinoff success: The role of knowledge intermediaries within an entrepreneurial university ecosystem. *Small Business Economics* 47 (3), 633–656. doi:10.1007/s11187-016-9756-3.

Hoerber, T., Weber, G., 2021. Introduction – The European environmental conscience in EU politics 1, in: Hoerber, T., Weber, G. (Eds), *The European environmental conscience in EU politics*. Routledge, pp. 1–23.

Hojnik, J., Ruzzier, M., 2016. What drives eco-innovation? A review of an emerging literature. *Environmental Innovation and Societal Transitions* 19, 31–41. doi:10.1016/j. eist.2015.09.006

Horner, S., Jayawarna, D., Giordano, B., Jones, O., 2019. Strategic choice in universities: Managerial agency and effective technology transfer. *Research Policy* 48 (5), 1297–1309. doi:10.1016/j.respol.2019.01.015

Hovi, J., Sprinz, D.F., Sælen, H., Underdal, A., 2016. Climate change mitigation: A role for climate clubs? *Palgrave Communications* 2 (1). doi:10.1057/palcomms.2016.20

Hovi, J., Sprinz, D.F., Sælen, H., Underdal, A., 2019. The club approach: A gateway to effective climate co-operation? *British Journal of Political Science* 49 (3), 1071–1096. doi:10.1017/S0007123416000788

Howells, J., 2006. Intermediation and the role of intermediaries in innovation. *Research Policy* 35 (5), 715–728. doi:10.1016/j.respol.2006.03.005

Klewitz, J., Hansen, E.G., 2014. Sustainability-oriented innovation of SMEs: A systematic review. *Journal of Cleaner Production* 65, 57–75. doi:10.1016/j.jclepro.2013.07.017

Levin, K., Cashore, B., Bernstein, S., Auld, G., 2012. Overcoming the tragedy of super wicked problems: Constraining our future selves to ameliorate global climate change. *Policy Sciences* 45 (2), 123–152. doi:10.1007/s11077-012-9151-0

Lockett, A., Wright, M., Franklin, S., 2003. Technology transfer and universities' spin-out strategies. *Small Business Economics* 20 (2), 185–200. doi:10.1023/A:1022220216972

Lozano, R., Bautista-Puig, N., Barreiro-Gen, M., 2022. Developing a sustainability competences paradigm in higher education or a white elephant? *Sustainable Development* 30 (5), 870–883. doi:10.1002/sd.2286

Mayring, P., 2015. *Qualitative Inhaltsanalyse: Grundlagen und Techniken* (12., überarb. Aufl.). Beltz.

Motta, M.J., 2018. Policy diffusion and directionality: Tracing early adoption of offshore wind policy. *Review of Policy Research* 35 (3), 398–421. doi:10.1111/ropr.12281

Ostrom, E., 1999. Coping with tragedies of the commons. *Annual Review of Political Science* 2 (1), 493–535. doi:10.1146/annurev.polisci.2.1.493

Owen-Smith, J., Powell, W.W., 2001. To patent or not: Faculty decisions and institutional success at technology transfer. *The Journal of Technology Transfer* 26 (1/2), 99–114. doi:10.1023/A:1007892413701

Platje, J., Kampen, R., 2016. Climate justice from a club good perspective. *International Journal of Climate Change Strategies and Management* 8 (4), 520–538. doi:10.1108/IJCCSM-11-2014-0131

Røpke, I., 2012. The unsustainable directionality of innovation – The example of the broadband transition. *Research Policy* 41 (9), 1631–1642. doi:10.1016/j.respol.2012.04.002.

Schot, J., Steinmueller, W.E., 2018. Three frames for innovation policy: R&D, systems of innovation and transformative change. *Research Policy* 47 (9), 1554–1567. doi:10.1016/j.respol.2018.08.011

Schwartz, H.M., 2017. Club goods, intellectual property rights, and profitability in the information economy. *Business & Politics* 19 (2), 191–214. doi:10.1017/bap.2016.11

Siegel, D.S., Waldman, D., Link, A., 2003. Assessing the impact of organizational practices on the relative productivity of university technology transfer offices: an exploratory study. *Research Policy* 32 (1), 27–48. 10.1016/S0048-7333(01)00196-2

Smink, M., Negro, S.O., Niesten, E., Hekkert, M.P., 2015. How mismatching institutional logics hinder niche–regime interaction and how boundary spanners intervene. *Technological Forecasting and Social Change* 100, 225–237. doi:10.1016/j.techfore.2015.07.004

Smith, A., 2007. Translating sustainabilities between green niches and socio-technical regimes. *Technology Analysis & Strategic Management* 19 (4), 427–450. doi:10.1080/09537320701403334

Sprinz, D.F., Sælen, H., Underdal, A., Hovi, J., 2018. The effectiveness of climate clubs under Donald Trump. *Climate Policy* 18 (7), 828–838. doi:10.1080/14693062.2017.1410090

Stahel, W.R., 2016. The circular economy. *Nature* 531 (7595), 435–438. doi:10.1038/531435a

Stephens, J.C., Hernandez, M.E., Román, M., Graham, A.C., Scholz, R.W., 2008. Higher education as a change agent for sustainability in different cultures and contexts. *International Journal of Sustainability in Higher Education* 9 (3), 317–338. doi:10.1108/14676370810885916

Swamidass, P.M., 2013. University startups as a commercialization alternative: Lessons from three contrasting case studies. *The Journal of Technology Transfer* 38 (6), 788–808. doi:10.1007/s10961-012-9267-6

Tether, Bruce S., Tajar, Abdelouahid, 2008. Beyond industry–university links: Sourcing knowledge for innovation from consultants, private research organisations and the public science-base. *Research Policy* 37 (6), 1079–1095.

Theodoraki, C., Messeghem, K., Rice, M.P., 2018. A social capital approach to the development of sustainable entrepreneurial ecosystems: An explorative study. *Small Business Economics* 51 (1), 153–170. doi:10.1007/s11187-017-9924-0

Tiba, S., van Rijnsoever, F., Hekkert, M.P., 2020. The lighthouse effect: How successful entrepreneurs influence the sustainability-orientation of entrepreneurial ecosystems. *Journal of Cleaner Production* 121616. doi:10.1016/j.jclepro.2020.121616

Tiemann, I., Fichter, K., Geier, J., 2018. University support systems for sustainable entrepreneurship: Insights from explorative case studies. *International Journal of Entrepreneurial Venturing* 10 (1), 83. doi:10.1504/IJEV.2018.090983

Vinig, T., Lips, D., 2015. Measuring the performance of university technology transfer using meta data approach: The case of Dutch universities. *The Journal of Technology Transfer* 40 (6), 1034–1049. doi:10.1007/s10961-014-9389-0

Volkmann, C., Fichter, K., Klofsten, M., Audretsch, D.B., 2021. Sustainable entrepreneurial ecosystems: An emerging field of research. *Small Business Economics* 56 (3), 1047–1055. doi:10.1007/s11187-019-00253-7

Wagner, M., Schaltegger, S., Hansen, E.G., Fichter, K., 2019. University-linked programmes for sustainable entrepreneurship and regional development: How and with what impact? *Small Business Economics* 56, 1141–1158. doi:10.1007/s11187-019-00280-4

Wakkee, I., van der Sijde, P., Vaupell, C., Ghuman, K., 2019. The university's role in sustainable development: Activating entrepreneurial scholars as agents of change. *Technological Forecasting and Social Change* 141, 195–205. doi:10.1016/j.techfore.2018.10.013

Yin, R.K., 2003. *Case study research: Design and methods*. Thousand Oaks.

Zhou, R., Tang, P., 2020. The role of university knowledge transfer offices: Not just commercialize research outputs! *Technovation* 90-91, 102100. doi:10.1016/j.technovation.2019.102100

10 The Long Walk to Digital Sobriety in the European Union

A French Case Study

Maxence Tétard and Marjorie Tendero

Introduction

Information technologies (IT) are omnipresent in our daily lives, both personal and professional. They encompass a broad range of tools, systems, and processes that are employed to manage, store, transmit, or retrieve information. They include, among others, computers and computing devices, 3D printing, artificial intelligence (AI), cloud computing, robotics, and the Internet of Things.

The environmental consequences of digital technologies are now widely acknowledged (ADEME, 2017; Bordage, 2019). Digital technologies are responsible for nearly 4% of worldwide carbon emissions, exceeding those generated by the civil aviation industry (Efoui-Hess, 2019). This proportion could reach 8% of total greenhouse gas (GHG) emissions by 2025, according to projections developed by Andrae and Edler (2015). The energy consumption of IT is increasing by 9% every year (The Shift Project, 2019). This consumption has doubled since 2013, along with its emissions. No other sector has experienced such intense growth (Flipo, 2022).

The Paris Agreement and the Sustainable Development Goals, especially goals 7 and 12, have emphasised the importance of GHG reduction. At the EU level, the European Climate Law sets the legal framework for achieving climate neutrality by 2050 (European Climate Law, 2021). The EU is pursuing the objective of achieving a 55% net reduction of GHG by 2030. As IT constitute a significant proportion of these emissions, there is a growing call to embrace digital sobriety (Ferreboeuf, 2019, 2022; Flipo, 2021). Indeed, to limit global warming to a maximum of 1.5°C, the IT sector must decrease its emissions by more than 40 % by 2030 (Freitag et al., 2021). Thus, initiatives undertaken by the European Union (EU) aim to incorporate environmental concerns into the energy realm. According to Hörber (2012), this constitutes the first criterion for assessing EU actions as indicative of an environmental conscience in Europe (Hörber, 2012). His framework adds three additional criteria for identifying whether specific development related to the EU's policies reflect environmental consciousness: the innovative nature of solutions or a commitment to progress; the European Commission's (EC) expertise and therefore legitimacy for political leadership; and the grass-roots support and demand for

DOI: 10.4324/9781032656359-14

effective solutions. This chapter aims to analyse the development of the digital sobriety at the EU level, with a focus on French digital service companies (DSC). Our objective is to analyse whether there is a digital environmental conscience in Europe by assessing the different selection criteria defined by Hörber (2012). At the microeconomic level, our objective is to examine how DSCs can promote digital sobriety within the regulatory framework of the EU?

Our chapter is organised as follows: Section 2 provides an in-depth exploration of the concept of digital sobriety. Section 3 shows the evolution of the regulatory framework concerning digital transition, corresponding to the second criterion of Hörber's (2012). In section 4, we delve into the methodology used for data collection and analysis and present our case study. The main findings derived from our case study are presented in section 5 and discussed. Finally, a conclusion about the development of the digital sobriety conscience at the EU level is presented in section 6.

The arc of sobriety: Historical origins and contemporary challenges

This section examines the concept of sobriety using philosophical and historical approaches (A). Then, it also considers its contemporary application in the realm of digital technology (B).

The concept of sobriety

The concept of sobriety has its roots in Antiquity. Greek and Roman philosophers such as Aristotle, Epicurus, and Seneca advocated for frugality, moderation, or temperance as a means to attain happiness (Cézard & Mourad, 2019). The concept of sobriety as a common spiritual value can also be found in different religions such as Christianity, Buddhism, Islam, and Judaism. These religions share a common emphasis on renouncing material wealth to achieve a better way of living, although different practices may be observed within each religion (Gregg, 1936).

In 1936, the American social philosopher Richard Bartlett Gregg (1885–1974) published his book entitled: *The Value of Voluntary Simplicity*. He highlighted the benefits associated with this type of lifestyle:

> Voluntary simplicity involves both inner and outer conditions. It means singleness of purpose, sincerity, and honesty within, as well as avoidance of exterior clutter, of many possessions irrelevant to the chief purpose of life. It means an ordering and guiding of our energy and our desires, a partial restraint in some directions to secure greater abundance of life in other directions. It involves a deliberate organization of life for a purpose.
>
> (Gregg, 1936, p. 5)

Even though the word "sobriety" is not employed in his book, Gregg advocated for a shift towards more conscious and more responsible consumption, as well as the adoption of sustainable production practices. His book can be considered as the beginning of the critique of consumerism that later emerged in Europe during the 1960s. These reflections were further explored by movements related to degrowth. Degrowth movements aim to focus on essential uses and consumption, as well as to reflect on new indicators of development and wealth that are not solely based on economic growth (Karakas, 2023; Missemer, 2017).

Degrowth is linked with the concept of sobriety: both concepts challenge the dominant paradigm of a continuous economic growth and emphasise the need to prioritise essential needs over excessive material desires. However, the objectives between these two concepts are different. The aim of degrowth movements is to move away from the current orthodox model of perpetual economic growth, while the aim of sobriety is to promote a more balanced way of life (Demaria et al., 2013). Besides, degrowth focuses on the macroeconomic level while sobriety focuses on a more microeconomic level as it concerns individuals' lifestyles (such as voluntary simplicity, minimalism, zero waste, and self-sufficiency), and choices such as reducing consumption by recycling. That is the reason why the concept of sobriety is sometimes translated with the word "sufficiency". This concept refers to having enough to meet one's needs. In a sustainability context, sufficiency can imply consuming only what is necessary (Alcott, 2008).

According to Guillard (2021), sobriety is a lifestyle that involves not just consuming better but also, and critically consuming less. This lifestyle can be expressed in consumption of energy and digital technologies as well as via material objects. The notion of sobriety includes not only the idea of consuming less, but also encompasses broader ethical dimensions that require a shift towards sustainable and ethical consumption practices.

The concept of digital sobriety

The concept of sobriety was first applied in the energy domain. In France, the association Négawatt, that gathered energy professionals, experts, and citizens, has developed one of the most comprehensive definitions of the concept of sobriety. Négawatt defines sobriety as "a process of reducing superfluous consumption through a prioritisation of needs that can be applied at the individual level" (Chatelin 2016, p. 11). Négawatt's conceptualisation of the sobriety aligns with the tenet of sufficiency as both definitions emphasise discernment regarding essential requisites. For example, sobriety involves not possessing several devices that provide analogous functionalities (such as a computer or tablet), but having just one, ideally a reconditioned purchase, that is used to meet a need rather than to stave off boredom (Guillard, 2021). The concept of digital sobriety refers to the conscious and moderate use of digital technologies with the aim of reducing their environmental footprint.

The concept of digital sobriety goes further than the concept of Green IT. Green IT seeks ways to achieve environmental optimisations of technological infrastructure, such as enhancing server energy efficiency. The notion of digital sobriety aims to reduce IT uses that are excessive and detrimental to the environment (Péréa et al., 2023). In other words, whereas Green IT poses the question "how can we render this technology more environmentally compliant?", digital sobriety wonders "is this technology truly indispensable?" or "is it possible to use it less?".

The unprecedented onset of the Covid-19 pandemic catalysed a massive shift towards digital platforms to facilitate various aspects of daily life, ranging from professional engagements, such as remote working, to personal endeavours like social interactions. This seismic shift to the digital realm has augmented the energy consumption of data centres, networks, and electronic devices. This dependency was further accentuated by the geopolitical turmoil in Ukraine, which exacerbated concerns about the sustainability and the security of energy supply chains. Consequently, this crisis has intensified the scrutiny on the sustainability of digital consumption patterns.

Environmental and digital shifts in the European Union

1960–2000: The beginning of the environmental transition

We can observe growing public awareness of environmental issues at the EU level. However, from the 1970s to the early 1980s, the European environmental conscience was still in its infancy (Tendero, 2021). Thus, the process of the EU's full commitment to sustainability was rather slow. Indeed, at that time, environmental benefits were considered to be only a by-product of the need for energy saving (Hörber, 2012).

The 1987 Single European Act marked the first time that the EU included "environmental protection integration" into its policies, which demanded a cross-check of all sectoral policies with respect to environmental goals.

The ecological transition has been part of the public policy agenda of the EU since the 1990s. The Maastricht Treaty (1992) formally recognised the importance of environmental protection and sustainable development. Environmental protection was incorporated into the core objectives of the EU. The Fifth Environmental Action Programme (EAP) (1992–2000) also marked a significant shift in the EU's environmental policy, as it integrated environmental concerns into other EU policies and emphasised the importance of economic instruments and voluntary agreements, for example. This EAP includes, for instance, the Integrated Pollution Prevention and Control Directive (1996/61), the Ambient Air Quality Directive (96/62), and the Water Framework Directive (2000/60). Then, the Amsterdam Treaty (1997) introduced a specific article that called for clear consideration of environmental principles in all EU actions and policies (Barnes & Hörber, 2013).

These first policies at the EU level paved the way for sustainability and environmental consciousness in the EU, the specific intersection of ecology and the digital realm became more pronounced in the 2000s. Three major European directives were adopted which focused on digital products and their environmental footprint in order to promote sustainability, energy efficiency and environmentally conscious usage (Flipo, 2022): the Energy using Products (EuP) Directive (2005/32/EC), the Restriction of Hazardous Substances (RoHS) Directive (2011/65/EU), and the Waste Electrical and Electronic Equipment (WEEE) Directive (2012/19/EU).

2000–2019: The beginning of the digital transition in Europe

First, the EuP Directive aimed to establish a framework for enhancing the energy efficiency of energy-consuming products throughout their life cycle. This directive imposed specific requirements on manufacturers regarding the energy efficiency and environmental performance of their products. However, these requirements were defined at the product level (Hinchliffe & Akkerman, 2017). That is why the Eco-design Directive (2009/125/EC) expanded the scope of regulation to include a wider variety of products and established minimum energy efficiency standards for those products. Furthermore, the Energy Labelling Directive (2010/30/EU) focused on energy labelling of products to inform consumers about their relative energy efficiency. This Directive required that products be labelled with a standardised energy label, ranging from A (very efficient) to G (inefficient), to assist consumers in making informed decisions regarding energy efficiency. Besides, under the Ecodesign Directive, there have been efforts to establish minimum repairability requirements for certain product categories. In this field, we are also observing a movement in Europe called "Right to Repair", which aims to provide consumers with repair services access (Svensson et al., 2018).

Second, the RoHS Directive restricts the use of certain hazardous substances, such as lead, mercury, and cadmium, for instance, in electrical and electronic equipment (Selin & VanDeveer, 2006). The primary objective was to minimise the environmental and health risks associated with the production and disposal of electronic products. Thus, this directive encouraged the adoption of cleaner and more environmentally friendly production practices within the electronics industry. This transition towards cleaner manufacturing processes not only reduces the ecological footprint of digital devices, but also contributes to the overarching environmental sustainability objectives of the EU.

In the context of regulations related to the digital transition, the WEEE Directive played a crucial role. This directive addressed the management of waste arising from electrical and electronic equipment. It imposed the concept of producer responsibility for the recycling and disposal of products at the end of their useful life. This directive is aligned with the principles of the circular economy and seeks to reduce the environmental impact of electronic waste, which is an integral part of the digital transition.

These directives have promoted the development of digital devices that consume fewer resources and have a longer lifespan, which corresponds to digital sobriety. However, in all these policies, digital technologies were implicitly and indirectly addressed.

2019–Present: Advancing sustainability through Digitalisation

Since 2019 and into the early 2020s, several European policies have become intertwined with digitalisation as they aim to facilitate a transition towards a forward-thinking and environmentally sustainable society. Indeed, in 2019, the EC adopted the European Green Deal with the objective to reduce the net emissions of GHG to zero in the EU by 2050. Digital technologies are explicitly addressed: "the EU should also promote and invest in the necessary digital transformation and tools as these are essential enablers of the changes" (European Commission, 2020b).

The European Green Deal defines a long-term strategy for a sustainable Europe through two main initiatives that can be related to digital sobriety: the New Circular Economy Action Plan and the planned Zero Pollution Action Plan (Bahn-Walkowiak et al., 2020). The New Circular Economy Action Plan, released in early 2020, states that:

> Building on the single market and the potential of digital technologies, the circular economy can strengthen the EU's industrial base and foster business creation and entrepreneurship among SMEs. Innovative models based on a closer relationship with customers, mass customisation, the sharing and collaborative economy, and powered by digital technologies […] will not only accelerate circularity, but also the dematerialisation of our economy and make Europe less dependent on primary materials.
>
> (European Commission, 2020a)

In 2020, the EC introduced three strategic initiatives: the new Digital Strategy, the New Industrial Strategy[1], and the European SMEs Strategy[2]. In February 2020, the EC adopted its new Digital Strategy, outlining a vision for a transition to a healthy planet and a new digital world. The European Data Strategy and the White Paper on Artificial Intelligence constituted the two main pillars of this Digital Strategy in Europe. Both documents acknowledged the environmental impacts and detrimental effects of the digitalisation. However, digitalisation is also recognised as a way to address environmental emergencies, combat environmental degradation, and climate change: "to develop an ambitious policy agenda for using digital solutions to achieve the zero pollution ambition" (European Commission, 2021). The New Industrial strategy, launched in March 2020 by the EC, aims to strengthen EU's industrial sector. It emphasises the importance of sustainability and digitalisation in driving industrial competitiveness. The importance of advanced digital technologies is recognised as essential to improve the efficiency of industrial processes and reduce their environmental impacts. In the same month, the EC launched the SMEs strategy. It is a targeted initiative focused on empowering SMEs within the EU. Its primary objective is to enable SMEs to flourish in a digital and sustainable landscape. This strategy defines support measures for SMEs to adopt digital technologies to enhance their competitiveness and their transition toward greener practices.

In 2023, the EU has continued to advance its regulatory framework for digital technologies with the adoption of the AI Act. This landmark legislation aims to ensure the safe and ethical development and use of AI. By addressing the ethical and environmental implications of AI, the AI Act complements the broader objectives of the Digital Strategy and reinforces the EU's dedication to sustainable and responsible technological advancement.

These strategies recognised the pivotal role of digitalisation in achieving environmental objectives and economic resilience. Even if they are indirectly linked with the digital sobriety, they share common objectives by promoting resource efficiency and responsible digitalisation.

Case study and methodologies

In this section, we provide information about our case study: Nell'Armonia, a French DSC (A). We also describe the methodologies used in this chapter (B), which includes semi-directive interviews and secondary data analysis.

A French digital service company: Nell'Armonia

Defining DSCs in the EU lacks a universally accepted framework. Therefore, various EU and international institutions have developed definitions and classifications for companies operating within the digital services sector. In France, the digital sector is typically categorised based on international sectoral classifications (ISIC, revision 4). It encompasses two primary components: the digital goods sector, which includes the manufacturing of computers, electronic, and optical products (sector 26), and the digital services sector, including software publishing services (sector 582), telecommunications services (sector 61), and computer and digital engineering services (sectors 62–63). In 2015, there were approximately 3,000 companies operating in this sector, employing 142,570 individuals (Gaglio & Guillou, 2018). For this reason, we consider that DSCs encompass a wide range of businesses, ranging from e-commerce platforms and online marketplaces to digital marketing agencies and software development firms. These companies leverage digital technologies to deliver various services, including software development, data analytics, digital marketing, and cloud computing.

Nell'Armonia was a consulting and technology firm, based in Levallois-Perret, and founded in 2005, that stands among the leaders in Enterprise Performance Management (EPM) solutions ranging from consultancy to implementation and maintenance. In 2020, Nell'Armonia reported a turnover of €17,660,000. In 2021, Accenture acquired Nell'Armonia. This acquisition was a strategic move for Accenture, as it expanded the company's EPM expertise. At the time of the acquisition, Nell'Armonia employed over 135 employees[3]. According to the annual ranking of DSCs in 2021, Accenture ranked third (€2.3 billion), following Capgemini (€4.2 billion) and SCC France (€2.6 billion).

Methodologies

To understand the role that DSCs can play in advancing digital sobriety in France, we adopted mixed methods research that combined qualitative and quantitative techniques.

Semi-structured interviews

In our research, we collected our data using semi-structured interviews. They are considered a suitable method when the researcher aims to gain a comprehensive understanding of the individual participants' unique perspectives. One significant advantage of semi-structured interviews is their ability to maintain a clear focus during the interview while allowing for the exploration of emerging relevant ideas (Moser & Korstjens, 2017). This flexibility enabled a deeper understanding of the concept of digital sobriety. We developed an interview questionnaire designed to explore the considerations of digital impacts, both internally and externally, with a focus on the context of the company's client relationships. The interview guide included questions regarding the eco-environmental practices implemented within the organisation and practices aimed at optimising digital resources (see appendix A).

These interviews involved eight individuals selected for their expertise in the fields of digital technology and/or sustainable development (refer to Appendix B for the list of participants involved in the semi-structured interviews). Participants were also selected considering their hierarchical positions within the value chain of Nell'Armonia. Interviewees included CSR managers, IT managers, and business partners from the firm, resulting in a sample of five experts in digital sobriety (interviewees 1, 2, 3, 5 and 8) and three individuals employed by Nell'Armonia (interviewees 4, 6 and 7), including the CTO and the CSR project manager of the firm. A balance was achieved between individuals with the greatest knowledge of the subject matter and those offering different perspectives (Guest et al., 2013). These interviews lasted from 30 to 60 minutes, with an average duration of 49 minutes. The interviews were audio-recorded as we obtained the free consent of all our participants. In total, 6 hours and 30 minutes of interviews were transcribed. Coding and themes identification were performed using Taguette, which is a free and open-source qualitative data analysis software (Rampin & Rampin, 2021). To analyse the data, each interview was considered independently, which facilitated the identification of specific themes and the conduction of a comparative analysis across interviews.

Secondary data analysis from France Num Barometers

Second, we used secondary data analysis from France Num Barometers to investigate the perceptions and uses of digital technologies by SMEs in France[4]. Secondary data analysis involves re-examining existing datasets (primary data) to address new research inquiries (Doolan & Froelicher, 2009). This method is widely used in social sciences (Johnston, 2017). It provides access to high-quality datasets characterised by substantial sample sizes and well-considered sampling strategies, enhancing validity and generalisability of the results (Smith et al., 2011).

France Num Barometers are annual surveys conducted by the French government agency responsible for business development since 2021. These surveys are conducted among a representative sample of French SMEs (see appendix C for an overview of sample sizes and questionnaire format). The surveys cover a wide range of topics, including the use of digital technology for business operations, for marketing and sales, as well as for employee communication and collaboration. These surveys are considered as an important tool for understanding the digital maturity of French SMEs and the challenges and opportunities they face. Moreover, they enabled us to establish a valuable connection with the research conducted through semi-structured interviews, contributing to a holistic understanding of the digitalisation process and its impact across the diverse stakeholders surveyed within the value chain.

Results

We present the thematic results obtained from the analysis of semi-structured interviews and discuss them considering the results from the France Num Barometers. The analysis of semi-structured interviews enabled us to identify four main thematic areas: the sustainability of equipment lifespan and usage (A), the limitation of consumption of digital goods and services (B), the pursuit of a balance between sobriety, costs, and performance (C), and the promotion of awareness about the environmental impacts of digital technologies (D).

The sustainability of equipment lifespan and usage

A consistent theme is the emphasis on maximising the lifespan of hardware components. Interviewees recognise that a significant portion of the environmental footprint of IT is associated with the manufacturing and disposal of physical hardware.

> In fact, a significant majority (80%) of the environmental footprint of IT is related to the construction/manufacturing of hardware. The challenge is, therefore, to maximize the lifespan of IT infrastructure (hardware) as much as possible.
>
> (Interviewee 1, 22 February 2021)

> The most significant impact of digital technology on the environment is the manufacturing of hardware, all physical components. Therefore, the challenge is to maximise the lifespan of hardware as much as possible. This also applies to the physical infrastructure of the cloud, such as data centers and other facilities.
>
> (Interviewee 2, 24 February 2021)

For this reason, it seems there is a call for eco-design IT projects to avoid frequent hardware changes. Repairing and refurbishing hardware are mentioned as practices to prolong the lifecycle of machines, which aligns with the principles of digital sobriety.

> Personally, I want to keep my PC for as long as possible, and I have already refused several times to have it replaced. As long as it does what I need for my job, I'm perfectly fine with it.
>
> (Interviewee 4, 3 March 2021)

> When launching an IT project, it's essential to ensure that digital development and architectural services are eco-designed to avoid changing hardware versions. Indeed, the most significant impact lies in the purchase (and thus the manufacturing, which requires the extraction of minerals and rare earths) of computer hardware (everything physical). The goal is to extend the life of this hardware for as long as possible. The act of purchase should be postponed as much as possible. Instead, it is sometimes possible to repair or opt for refurbished options.
>
> (Interviewee 5, 4 March 2021)

While energy efficiency is deemed as important, simply replacing hardware with newer, more energy-efficient models may not always be the best solution. The primary goal is to make the most of existing hardware and infrastructure.

> The objective is not just to focus on energy consumption. For example, it may not necessarily be the solution to buy a newer server that consumes less energy. The goal is always to maximise the lifespan of existing hardware.
>
> (Interviewee 1, 22 February 2021)

Having in-house skills for repairing workstations and collaborating with suppliers to extend the life of machine is a proactive approach to achieve digital sobriety.

> We also have good in-house skills to repair workstations and interact with suppliers when necessary to ensure a long lifecycle for our machines.
>
> (Interviewee 7, 24 March 2021)

The limitation of the consumption of digital goods and services

The insights gleaned from the second category of verbatims shed light on another key aspect of digital sobriety. As most environmental impacts in digital technology are concentrated in the manufacturing of hardware important, it is also crucial to extend the lifespan of existing hardware. Additionally, the verbatims emphasise the need for optimising resource usage in cloud environments.

I can tell you that sometimes development environments run 24 hours a day in the cloud, even though they are only used for 3 hours.

> (Interviewee 1, 22 February 2021)

Indeed, keeping development environments running continuously, even when they are not actively used, contributes to unnecessary resource consumption. Besides, the role of Digital Service Companies is to minimise computing power. This design aspect plays a pivotal role in promoting digital sobriety by ensuring that digital services are eco-friendly.

The challenge for IT service companies (ESNs) is to design IT services that consume the least computing power possible.

> (Interviewee 2, 24 February 2021)

The vast majority of the environmental impacts of digital technology are concentrated in the manufacturing of hardware equipment. Therefore, the goal is to maximise the lifespan of these infrastructures. Electricity consumption and data volume are secondary concerns.

> (Interviewee 3, 1 March 2021)

Furthermore, the importance of managing data storage effectively is highlighted. This includes, for example, setting limits on email inbox sizes to encourage the deletion of unnecessary emails, thereby reducing the environmental impact associated with data storage.

It's up to the CIO to implement things like capping disk space, for example. When I talk about disk space, I also mean email inboxes. If the space is limited, employees will naturally develop the habit of deleting the heaviest emails, with large attachments, over time.

> (Interviewee 3, 1 March 2021)

I remember that back when I had limited storage space in my email inbox, I established a routine of deleting the largest and oldest emails. The technical constraint helped limit storage space and thus the environmental impact. We should gradually reintroduce this constraint and raise awareness among employees on this issue. It could take the form of a common routine that we all do at the same time.

> (Interviewee 4, 3 March 2021)

The role of billing structure as a potential incentive for organisations to practice data optimisation and retention of essential data is important. It encourages organisations to be more mindful of their data storage practices, not only from a cost-efficiency perspective, but also as a means to indirectly promote digital sobriety.

> That's an interesting point about Anaplan. Being billed based on the amount of data in the solution can be a strong incentive for organisations to optimise and retain only essential data. This approach not only aligns with cost-efficiency but also indirectly promotes digital sobriety by encouraging organisations to minimise unnecessary data storage, which can contribute to reduced energy consumption and a smaller environmental footprint.
>
> (Interviewee 7, 24 March 2021)

Recommendations for more environmentally friendly email practices are also shared.

> Sending fewer emails isn't necessarily the solution. The key is to send better emails, meaning limiting the number of recipients to genuine needs, opting for face-to-face communication if the person is nearby, and reducing the size of large attachments. Additionally, the duration of storage for old, large emails on servers is also crucial. Here are some best practices recommended by ADEME (French Agency for Ecological Transition): regularly clean your mailbox; unsubscribe from unnecessary newsletters. Compress the size of attachments or use temporary deposit sites precisely target your recipients.; create a signature without images or logos, at least for internal exchanges.
>
> (Interviewee 5, 4 March 2021)

However, data management within collaborative tools, such as Teams and SharePoint, can present challenges in terms of organising and deleting files.

> On the other hand, sorting and organising files in Teams and SharePoint are not facilitated: deleting a team file in SharePoint is more challenging than a personal file. Cleaning up in SharePoint is more complex, as sometimes there is a need for collaboration to delete files.
>
> (Interviewee 8, 25 March 2021)

The use of Teams to reduce email-related environmental impacts is emphasised by some interviewees. This vision is mitigated as Teams can be harnessed to implement environmentally friendly digital practices. These insights underscore the significance of developing effective metrics and measurement systems to comprehensively assess and monitor the environmental impacts of digital practices, ensuring that efforts toward digital sobriety are quantifiable and can be improved.

> Our solutions were designed to centralise information in one place to avoid budgetary processes involving endless email exchanges (budgets in large Excel spreadsheets). So indirectly, we are still reducing the environmental impact of emails.
>
> (Interviewee 4, 3 March 2021)

> I find that Teams is a great way to implement more environmentally friendly digital practices.
>
> (Interviewee 7, 24 March 2021)

> Microsoft 365 cloud isn't necessarily very environmentally friendly. It offers massive storage space, especially for emails, which can be quite resource intensive.
>
> (Interviewee 6, 5 March 2021)

The pursuit of a balance between sobriety, costs, and performance

Interviewees acknowledge that, in many IT projects, the budget tends to take precedence over environmental considerations. They highlight the challenges of introducing responsible digital practices in projects primarily driven by financial constraints.

> The environmental criteria are often not the priority in an IT project. However, it is important to try to find a balance with the financial constraint.
>
> (Interviewee 3, 1 March 2021)

The participants emphasise the importance of striking a balance between environmental considerations and meeting client requirements.

> There is a trade-off to be made with the client. The goal is to try to show them that a more frugal information system will be less energy-consuming, which will ultimately lead to energy savings and cost savings.
>
> (Interviewee 1, 22 February 2021)

Besides, clients may prioritise factors like security, which can sometimes lead to the need for multiple infrastructures. This highlights the challenges of aligning environmental sustainability with client demands.

> Depending on the client's requirements, you may need multiple infrastructures, even for simple security reasons. This often requires storing data in two different locations. It is necessary to find a balance between security, availability, and environmental footprint.
>
> (Interviewee 1, 22 February 2021)

However, by showcasing how these practices can lead to cost savings (e.g., reduced infrastructure renewal and energy consumption), organisations can make a more compelling case for adopting environmentally friendly approaches.

> It's also important to demonstrate the ROI of more responsible digital practices to project decision-makers. Although decisions have mostly been budget-driven until now, the key is to convince them that

implementing less resource-intensive solutions not only reduces the environmental footprint but also reduces costs (infrastructure renewal, energy consumption, etc.).

(Interviewee 2, 24 February 2021)

One practical approach suggested by an interviewee is to present digital sobriety as an additional benefit rather than the main argument in discussions with clients.

When clients initiate IT projects, the budget often takes center stage, and it can be challenging to introduce the topic of responsible digital practices when their primary concerns are economic and technical. Making discussions about the environmental footprint of a digital project an additional benefit rather than the main argument is a pragmatic approach. By demonstrating that adopting digital sustainability measures can ultimately lead to cost savings, improved efficiency, and reduced long-term environmental impact, you can make it more appealing to clients who prioritize budget and technical concerns. In this way, environmental considerations become an attractive "cherry on top" of the overall project proposal.

(Interviewee 4, 24 March 2021)

Furthermore, some interviewees mention a level of dependence on partner publishers, whose priorities often focus on developing additional features and applications that may require more computing power. This dependence can influence the ability to prioritise environmental concerns.

We are somewhat dependent on our partner publishers. For now, their priority is more on developing additional features and applications, which require more computing power.

(Interviewee 4, 3 March 2021)

The promotion of awareness of the environmental impacts of digital technologies

The importance of raising awareness within organisations is highlighted in numerous interviews. This includes efforts to educate colleagues and IT departments about the environmental impacts of the digital technologies. Some even mention specific events like the Cyber World Cleanup Days as opportunities to engage employees in data clean-up activities.

You have various ways to raise awareness among your colleagues and your IT department internally. In March, the Cyber World Cleanup Day will be organised. This event was designed to encourage people to clean up their data for one day.

(Interviewee 1, 22 February 2021)

Interviewees stress the need to raise awareness among all types of employees.

What's important afterward is to raise awareness among all types of employees, not just developers or consultants.

(Interviewee 3, 1 March 2021)

The role of the users in managing data end-of-life is also recognised.

> User support in managing data end-of-life: deleting emails, transferring documents to the cloud, and improving file organisation for optimised management. The issue of dark data is significant. These are data that companies collect, process, and store during their regular activities. However, these data are never reused. In doubt, they are kept somewhere because it's unclear what to do with them. Some companies may not even be aware of their existence.
>
> (Interviewee 5, 4 March 2021)

However, promoting environmental awareness of the environmental impact of digital technologies within organisations may require management efforts. Long-time employees might find it challenging to adapt to new eco-friendly practices, but gradual change is seen as achievable.

> The only thing is that it requires some change management. People who have been working here for a long time are finding it difficult to adapt, but little by little, it will come.
>
> (Interviewee 7, 24 March 2021)

Discussions

According to the results from the France Num Barometers, SMEs leaders perceived digital technology as a real benefit for their businesses. In 2021, 78% of surveyed leaders held this view, and this percentage increased to 81% in 2022 (Direction Générale des Entreprises & CREDOC, 2021, 2022). However, in 2023, this positive perception stabilised slightly with 76% of leaders believing that digital technology represents a tangible benefit for their companies (Centre de Recherche pour l'Étude et l'Observations des Conditions de Vie, 2023). These perceptions seem coherent with the industrial and digital strategies developed at the EU level since 2020 that emphasise the positive role of the digitalisation to achieve the zero pollution ambition (European Commission, 2021).

However, the increasing use of AI in companies will present a significant challenge to digital sobriety. AI technologies require substantial server capacities, leading to higher energy consumption and a greater environmental impact. Therefore, robust infrastructure is needed to support AI. It can conflict with efforts to reduce digital footprints and optimise resource usage.

Surveyed leaders cited enhanced visibility and online sales as primary benefits of digital technology, with the latter being reported by 84% and 27% of respondents in 2023. Additionally, the adoption of collaborative tools, particularly instant messaging, has seen substantial growth, reaching 70% in 2023 compared to 33% in 2021 (Centre de Recherche pour l'Étude et l'Observation des Conditions de Vie, 2023; Direction Générale des Entreprises & CREDOC, 2021). This positive trend reflects the discussions in the interviews regarding the use of collaborative tools such as Teams and SharePoint.

Email usage and storage constitute another challenge for digital sobriety. While limiting email storage can help reduce environmental impact, legal constraints often require companies to store emails for extended periods for compliance and record-keeping purposes. This legal necessity can hinder efforts to minimise data storage and its associated environmental footprint. Another issue is the use of email signatures that include images, which increase the size of emails and thus their environmental impact. This practice is often dictated by company procedures and policies, making it another area where organisational change is needed to support digital sobriety.

There is also a growing concern about data security, with half of the leaders' expressing apprehensions about data safety. This concern was not expressed in our interviews.

Digital sobriety was not a specific focus in the 2021 and 2022 barometers, suggesting that it was not a prominent individual concern for business leaders. In 2023, a set of questions related to digital sobriety was posed to these leaders. Even if 42% of companies have taken actions in favour of digital sobriety, these actions remain limited (Centre de Recherche pour l'Étude et l'Observation des Conditions de Vie, 2023). Indeed, acquiring and/or improving IT equipment remains the top priority, followed by the desire to engage in social media communication and gain online visibility. Therefore, even though 81% of companies claim a preference for French or local (regional or departmental) service and software providers, these practices can potentially contradict digital sobriety efforts if the acquisition of new equipment takes precedence. Besides, this desire to engage in social media communication and increase online visibility can be counterproductive to digital sobriety, particularly when it involves intensive use of digital resources. Indeed, social media and online promotion often entail the creation of digital content, data storage and remote servers, as well as the streaming of videos and images, all of which consume energy and resources. Therefore, if these actions are not carried out responsibly, they can contribute to a larger digital footprint.

Furthermore, we observe disparities in the willingness to implement digital sobriety practices based on the age of the leaders (Centre de Recherche pour l'Étude et l'Observation des Conditions de Vie, 2023). Older leaders tend to be more hesitant to adopt such measures, confirming the hypothesis that adapting to changes is more complex for employees with greater seniority within the firm. The level of education of the leader also plays a role in the adoption of digitally sustainable practices: only 29% of leaders with educational attainment below a high school diploma are willing to implement such measures, compared to 46% for leaders with a bachelor's degree or higher (Centre de Recherche pour l'étude et l'observation des conditions de vie, 2023). Lastly, firm-related characteristics are also significant. Digital-focused companies, likely more aware of environmental pollution issues, are more supportive of implementing sustainable digital practices, with 56% of ICT (Information and Communication Technology) firms taking such actions, while in the agriculture or agri-food industry sectors, this proportion drops to approximately one-third. Finally, companies with 50 to 250 employees tend to undertake more actions in favour of digital sobriety than smaller companies (with fewer than 50 employees). This could be attributed to the presence of dedicated departments for Corporate Social Responsibility (CSR) and IT in such structures.

However, despite the growing awareness of digital sobriety and the EU's efforts to promote sustainable digital practices, a significant challenge remains in the reliance on US-based cloud and AI providers. Many companies still opt for services like OneDrive over European alternatives like Ionos or choose ChatGPT over Mistral AI, despite the potential environmental benefits of using local providers. This reliance on non-European providers not only raises concerns about data sovereignty and privacy but also has implications for digital sobriety. The data centres of these providers, often located outside Europe, may not adhere to the same environmental standards, or utilise renewable energy sources to the same extent as European providers. This can contribute to a larger carbon footprint and undermine efforts to achieve digital sobriety goals. Therefore, promoting the adoption of European cloud and AI providers could be a crucial step towards achieving digital sobriety within the EU.

Conclusion: A developing European digital sobriety?

Our objective was to analyse whether there is a digital environmental conscience in Europe by assessing the different selection criteria defined by Hörber (2012). This theoretical framework aims to discern whether specific developments related to the policies and actions of the EU mirror a broader environmental consciousness in Europe. They have distinguished between four selection criteria: the nexus between energy and environmental policies, the innovativeness of solutions or commitment to progress; the expertise of the EC and its political leadership legitimacy; and the grassroots support and demand for effective solutions. We started our study with a historical approach, delineating the evolution of environmental policies in the domain of digital sobriety. Our historical approach shows that the EU has embarked on a green and digital transition. These two dimensions are closely interrelated with the concept of digital sobriety. We demonstrate that three crucial phases have marked the evolution of environmental awareness regarding digital technology in the EU. These phases illustrate gradual shifts in the development of environmental consciousness within digital sobriety policies in the EU. The first phase, spanning from 1960 to 2000, was marked by the beginning of the environmental conscience in the EU. During this period, environmental concerns were primarily viewed as a by-product of energy-saving initiatives. The second phase, from 2000 to 2019, witnessed the convergence of the digital technologies and sustainability objectives. Since 2019, the link between digitalisation and sustainability is more explicit. The EU's commitment to reduce GHG emissions to zero and the recognition of digital technologies as tools to combat environmental challenges echo the first and the third criteria of the framework developed by Hörber (2012).

We also adopted a mixed methods approach to examine how digital sobriety can be achieved at a microeconomic level with a focus on DSCs. We conducted in-depth interviews to gain insights into the perspectives of experts and professionals in the digital and sustainability fields. We also used quantitative data from the France Num Barometers to provide a broader context for our findings. Our interviewees stressed the importance of extending IT hardware lifespan to reduce environmental impacts, focusing on repair, and eco-design. They also highlighted minimising new hardware purchases and optimising digital service use to enhance digital sobriety, emphasising data control and optimisation. Besides, committing to a sobriety-oriented journey entails a comprehensive re-evaluation of practices throughout the entire consumption process. Our interviews showed the need for raising awareness among employees and user support in managing data end-of-life. This process also prompts consumers to reexamine their connections with material possessions. While analysing the results from interviews and the France Num Barometers, it becomes evident that the fourth criterion outlined by Hörber and Weber (2022), which assesses grassroots support and demand for effective solutions, presents a nuanced perspective. The findings indicate that digital sobriety, while gaining some recognition, is still in its early stages as a prominent concern among business leaders. In the 2021 and 2022 barometers, digital sobriety did not emerge as a central focus, suggesting that it had yet to capture the attention of many decision-makers. In 2023, a set of questions directly related to digital sobriety was introduced, revealing that approximately 42% of companies have initiated actions in favour of digital sobriety. However, these endeavours, though promising, remain somewhat limited in scope. Indeed, the top priority still revolves around acquiring or upgrading IT equipment. Therefore, the digital sobriety journey in Europe is still in its infancy, with a long path ahead for achieving widespread commitment and comprehensive implementation.

Appendix

Interview guide

Introduction: Thank you very much for accepting my request for an interview and for your time. I am Maxence Tétard, I am a master's student at ESSCA School of Management. As part of my end-of-study work, I am carrying out a qualitative study on how a French digital service company can implement digital sobriety. I've prepared a few questions to help us keep things on track. Any information gathered will be used solely for the purpose of this research.

1 Presentation of the interviewee and their role in the company
2 Setting the context: Green IT and digital sobriety

3 Questions related to the environmental footprint of the firm (internal approach)

  - Questions about the hardware and its selection criteria: price, environmental score, obsolescence etc.
  - Questions about the software and its selection criteria: data centre and corporate website architecture.

4 Questions related to the environmental footprint for its partners and client (external approach)

  - Questions about the interest for the client to reduce the carbon footprint of the ICT system.
  - Questions about the communication efforts made by the firm to raise awareness of its commercial partners (including clients).
  - Questions about the eco-friendly practices.

5 Questions related to the broad use of digital resources.

  - Use of cloud computing and links with data management.
  - Use of web-based communication platforms such as SharePoint and Teams (Microsoft).
  - Use of remote working.
  - Other resource optimisation techniques.

6 Additional questions, if any, depending on the interviewee.

List of semi-structured interviews

#	Date	Function	Time spent in current position (Year)	Length of the interview (Minutes)
1	22/02/2021	CEO of a firm in green IT and Responsible Digital	4	30
2	24/02/2021	CEO of a firm in green IT and Responsible Digital	2	45
3	01/03/2021	CSR project manager	1	60
4	03/03/2021	Senior manager (Nell'Armonia)	15	60
5	04/03/2021	CSR Manager - Researcher	17	30
6	05/03/2021	Chief Technology Officer (Nell'Armonia)	11	45
7	24/03/2021	CSR Project manager (Nell'Armonia)	12	60
8	25/03/2021	Manager of Digital Sustainability and Data Architecture	5	60

Overviews of the France Num barometers

	2021	*2022*	*2023*
Sample size	2,796	4,671	9,453
Businesses with fewer than 10 employees	1,950	3,208 incl. 945 with zero employees	6,110 incl. 2,056 with zero employees
Survey period	19/03/2021–06/04/2021	25/02/2022–29/03/2022	09/03/2023–07/04/2023
Survey methods	Online: 2,371 Phone: 425 only for very small companies	Online: 3,848 Phone: 823	Online: 8,713 Phone: 740
Questionnaire structure	1 Company knowledge 2 Projects and motivations 3 Attitudes towards digital technology 4 Digital equipment level 5 Perceived costs and benefits to use digital technologies within the firm 6 Digital projects 7 Support for digital transformation	1 Company knowledge 2 Projects and motivations 3 Attitudes towards digital technology 4 Digital equipment level 5 Perceived costs and benefits to use digital technologies within the firm 6 Digital projects 7 Support for digital transformation	1 Company knowledge 2 Projects and motivations 3 Attitudes towards digital technology 4 Digital equipment level 5 Perceived costs and benefits to use digital technologies within the firm 6 Digital projects 7 Support for digital transformation 8 Sustainable development 9 Digital sovereignty

Data availability

Data will be made available upon request.

Notes

1 https://commission.europa.eu/strategy-and-policy/priorities-2019-2024/europe-fit-digital-age/european-industrial-strategy_en.
2 https://single-market-economy.ec.europa.eu/smes/sme-strategy_en.
3 https://newsroom.accenture.fr/fr/news/accenture-annonce-son-intention-acqu%C3%A9rir-nell-armonia-soci%C3%A9t%C3%A9-de-conseil-leader-dans-les-solutions-de-pilotage-de-la-performance-d-entreprise.htm.
4 https://data.economie.gouv.fr/pages/barometre-france-num/.

Bibliography

ADEME. (2017). *La Face cachée du numérique. Réduire les impacts du numérique sur l'environnement.* (Clés Pour Agir) [Guide]. ADEME.

Alcott, B. (2008). The sufficiency strategy: Would rich-world frugality lower environmental impact? *Ecological Economics, 64*(4), 770–786. https://doi.org/10.1016/j.ecolecon.2007.04.015

Andrae, A., & Edler, T. (2015). On global electricity usage of communication technology: Trends to 2030. *Challenges, 6*(1), 117–157. https://doi.org/10.3390/challe6010117

Bahn-Walkowiak, B., Magrini, C., Berg, H., Gözet, B., O'Brien, B., Arjomandi, T., Doranova, A., Le Gallou, M., Gionfra, S., Graf, V., Kong, M. A., Jordan, N., Miedzinski, M., & Bleischwitz, R. (2020). *Eco-Innovation and Digitalisatio. Case studies, environmental and policy lessons from EU member states for the EU green deal and the circular economy* [EIO Biennal Report]. Wuppertal Institute.

Barnes, P. M., & Hörber, T. C. (Eds.). (2013). *Sustainable development and governance in Europe: The evolution of the discourse on sustainability.* Routledge.

Bordage, F. (2019). *Empreinte environnmentale du numérique mondial* (Rapport d'études 2; p. 40). GreenIT.

Centre de Recherche pour l'étude et l'observation des conditions de vie. (2023). *Baromètre France Num. Résultats de l'enquête 2023* (Baromètre France Num, p. 61). Direction Générale des Entreprises.

Cézard, F., & Mourad, M. (2019). *Panorama sur la notion de sobriété: Définitions, mises en oeuvre, enjeux* (Rapport d'expertises, p. 52) [Rapport final]. ADEM.

Chatelin, S. (2016). *Qu'est-ce que la sobriété?* Fil d'Argent 5, pp. 11–13. Retrieved 2016-08-02 from http://www.negawatt.org/telechargement/Presse/1601_Fil-dargent_Qu-est-ce-que-la-sobriete.pdf

Demaria, F., Schneider, F., Sekulova, F., & Martinez-Alier, J. (2013). What is degrowth? From an activist slogan to a social movement. *Environmental Values, 22*(2), 191–215. https://doi.org/10.3197/096327113X13581561725194

Direction Générale des Entreprises & CREDOC. (2021). *Baromètre France Num 2021. Résultat de l'enquête 2021* (p. 60). Centre de Recherche pour l'étude et l'observation des conditions de vie.

Direction Générale des Entreprises & CREDOC. (2022). *Baromètre France Num 2022. Résultats de l'enquête 2022* (p. 60). Centre de Recherche pour l'étude et l'observation des conditions de vie.

Doolan, D. M., & Froelicher, E. S. (2009). Using an existing data set to answer new research questions: A methodological review. *Research and Theory for Nursing Practice, 23*(3), 203–215. https://doi.org/10.1891/1541-6577.23.3.203

Efoui-Hess, M. (2019). *Climate crisis: The unsustainable use of online video. The practical case study of online video* (p. 36). The Shift Project.

European Climate Law, Pub. L. No. 2021/119 (2021). http://data.europa.eu/eli/reg/2021/1119/oj

European Commission. (2020a). *A new circular economy action plan for a cleaner and more competitive Europe* [Communication from the Commission to the European Parliament, the Council, the European and Social Committee and the Committee of the regions]. European Commission.

European Commission. (2020b). *The European Green Deal. Europe's new growth strategy. A climate-neutral EU by 2050* (UNIDO Brussels Focus, p. 29). European Commission.

European Commission. (2021). *Digital solutions for zero pollution* (Commission Staff Working Document COM(2021) 400 final-SWD(2021) 141 final; Working Document, p. 50). European Commission.

Ferreboeuf, H. (2019). Pour une sobriété numérique. *Futuribles, 429*(2), 15. https://doi.org/10.3917/futur.429.0015

Ferreboeuf, H. (2022). Déployer la sobriété dans le numérique. *Revue Internationale et Stratégique, 128*(4), 127–133. https://doi.org/10.3917/ris.128.0127

Flipo, F. (2021). L'impératif de la sobriété numérique. *Cahiers Droit, Sciences & Technologies, 13*, 29–47. https://doi.org/10.4000/cdst.4182

Flipo, F. (2022). L'enjeu de la sobriété numérique: État des lieux en France. *Sens Public, 1635*, 35.

Freitag, C., Berners-Lee, M., Widdicks, K., Knowles, B., Blair, G. S., & Friday, A. (2021). The real climate and transformative impact of ICT: A critique of estimates, trends, and regulations. *Patterns, 2*(9), 100340. https://doi.org/10.1016/j.patter.2021.100340

Gaglio, C., & Guillou, S. (2018). Le tissu productif numérique en France. *OFCE Policy Brief, 36*, 1–19.

Golob, U., & Kronegger, L. (2019). Environmental consciousness of European consumers: A segmentation-based study. *Journal of Cleaner Production, 221*, 1–9. https://doi.org/10.1016/j.jclepro.2019.02.197

Gregg, R. B. (1936). *The value of voluntary simplicity* (2009th ed.). The Floating Press.

Guest, G., Namey, E. E., & Mitchell, M. L. (2013). Sampling in qualitative research. In *Collecting qualitative data. A field manual for applied research* (pp. 41–75). SAGE Publications.

Guillard, V. (2021). Towards a society of sobriety: Conditions for a change in consumer behavior. *Field Actions Science Reports, 23*(Special Issue), 36–39.

Hinchliffe, D., & Akkerman, F. (2017). Assessing the review process of EU Ecodesign regulations. *Journal of Cleaner Production, 168*, 1603–1613. https://doi.org/10.1016/j.jclepro.2017.03.091

Hörber, T. (2012). *The origins of energy and environmental policy in Europe* (0 ed.). Routledge. https://doi.org/10.4324/9780203083048

Hörber, T., & Weber, G. (2022). *The European environmental conscience in EU politics: A developing ideology.* Routledge.

Johnston, M. (2017). Secondary data analysis: A method of which the time has come. *Qualitative and Quantitative Methods in Libraries, 3*(3), 619–626.

Karakas, T. (2023). The degrowth movement in france: from the edges to the centre of the ecological debate. In E. Orofino & W. Allchorn (Eds.), *Routledge handbook of non-violent extremism: Groups, perspectives and new debates.* Routledge.

Kollmuss, A., & Agyeman, J. (2002). Mind the gap: Why do people act environmentally and what are the barriers to pro-environmental behavior? *Environmental Education Research, 8*(3), 239–260. https://doi.org/10.1080/13504620220145401

Missemer, A. (2017). Nicholas Georgescu-Roegen and degrowth. *The European Journal of the History of Economic Thought, 24*(3), 493–506. https://doi.org/10.1080/0967256 7.2016.1189945

Moser, A., & Korstjens, I. (2017). Series: Practical guidance to qualitative research. Part 1: Introduction. *European Journal of General Practice, 23*(1), 271–273. https://doi.org/10.1080/13814788.2017.1375093

Péréa, C., Gérard, J., & De Benedittis, J. (2023). Digital sobriety: From awareness of the negative impacts of IT usages to degrowth technology at work. *Technological Forecasting and Social Change, 194*, 122670. https://doi.org/10.1016/j.techfore.2023.122670

Rampin, R., & Rampin, V. (2021). Taguette: Open-source qualitative data analysis. *The Journal of Open Source Software, 6*(68), 1–5. https://doi.org/10.21105/joss.03522

Selin, H., & VanDeveer, S. D. (2006). Raising global standards: Hazardous substances and E-waste management in the European Union. *Environment: Science and Policy for Sustainable Development, 48*(10), 6–18. https://doi.org/10.3200/ENVT.48.10.6-18

Sharma, K., & Bansal, M. (2013). Environmental consciousness, its antecedents and behavioural outcomes. *Journal of Indian Business Research, 5*(3), 198–214. https://doi.org/10.1108/JIBR-10-2012-0080

Smith, A. K., Ayanian, J. Z., Covinsky, K. E., Landon, B. E., McCarthy, E. P., Wee, C. C., & Steinman, M. A. (2011). Conducting high-value secondary dataset analysis: An introductory guide and resources. *Journal of General Internal Medicine, 26*(8), 920–929. https://doi.org/10.1007/s11606-010-1621-5

Svensson, S., Richter, J. L., Maitre-Ekern, É., Pihlajarinne, T., Maigret, A., & Dalhammar, C. (2018). *The emerging 'right to repair' legislation in the EU and the U.S.* EventGoing Green Care Innovation. http://ci2018.care-electronics.net/

Tendero, M. (2021). The time has come! The development of the European environmental conscience. Evidence from the Eurobarometer surveys from 1974 to 2020. In T. C. Hörber & G. Weber (Eds.), *The European environmental conscience in EU politics* (1st ed., pp. 185–206). Routledge, Taylor & Francis Group.

The Shift Project. (2019). *Lean ICT - Towards digital sobriety* (p. 90) [Report of the working group directed by hugues Ferreboeuf]. The Shift Project.

Conclusion

Perspectives for the Governance of Sustainability in Europe and the European Environmental Conscience

Marjorie Tendero and Christoph Weber

The primary objective of this book was to provide a comprehensive examination of the existence, the evolution, and current state of environmental conscience within the EU. This was achieved by exploring the financial, innovative, and external policy dimensions that shape the EU's approach to sustainability. Through detailed analysis and case studies, this book aimed to highlight how the integration of environmental considerations into financial regulations, the promotion of technological and policy innovation, and the EU's leadership in international negotiations all contribute to a robust environmental conscience.

Hörber's theoretical framework, which posits the existence of a European environmental conscience, provides a valuable lens for understanding the integration of environmental considerations into various policy areas (Hörber, 2013; Hörber & Weber, 2022). This framework highlights the interconnectedness of energy and environmental policies, the innovative nature of solutions, the legitimacy of the European Commission's expertise, and grassroots support for effective solutions. These four elements, as defined by Hörber (2013), constitute the core components of the European environmental conscience and shape its development and implementation in EU policymaking.

The first dimension emphasises the intrinsic link between energy policies and environmental concerns. For example, the energy crisis of the 1970s highlighted the need to consider the environmental impacts of energy production and consumption (Matláry, 1997). This led to the integration of environmental considerations into European energy policies (Barnes & Hörber, 2015; Hörber, 2013). The second dimension highlights the importance of technological and policy innovation in addressing environmental challenges. The legitimacy of the European Commission's expertise in political leadership recognises its central role in the formulation and implementation of environmental policies. Its expertise and ability to mobilise Member States are essential for moving the European environmental agenda forward. The last dimension emphasises the importance of citizen engagement and civil society pressure in promoting ambitious environmental policies.

Hörber (2013) examines the emergence of an environmental protection agenda in relation to energy consumption in Britain, France, and Germany following the 1973 oil crisis. By analysing parliamentary discourses in the

DOI: 10.4324/9781032656359-15

energy sector, the author argues that a common European energy policy is the logical outcome of a general commitment to environmental protection across the EU. Then, building on Hörber's earlier work on the origins of the European environmental conscience, Hörber and Weber (2022) conducted a historical analysis of the development of the European environmental conscience, demonstrating its evolution from a marginal, energy-focused concern to a central ideology in European politics.

Applying and extending Hörber's theoretical framework, this book aims to inform policymakers, academics, and the public about the significant progress made by the EU in promoting sustainability and to provide insights into the ongoing challenges and future directions for environmental policy in Europe. Indeed, the comprehensive examination of environmental consciousness in Europe underlines a dynamic and multifaceted evolution towards sustainability. The chapters in this book have traced the evolution of this European Environmental Conscience from its early stages, driven by the need for energy provision and economic recovery, to its current manifestation in robust policies and innovative strategies aimed at achieving a sustainable future.

The environmental conscience in European finance

The first part of the book provides a comprehensive understanding of how sustainable finance and innovative financial instruments are central to the European Environmental Conscience. From theory to practice, these chapters illustrate how the EU is using these innovative financial instruments to align its economic policies with sustainability goals, thereby reinforcing its role as a global leader in sustainable finance and environmental protection.

Until recently, sustainability has not been a primary concern in financial regulation, which has traditionally focused on profit maximisation. The 2008 global financial crisis exposed the limitations of this approach and led to calls for greater government intervention. At the same time, the urgency of climate change prompted the integration of sustainability into financial regulation.

The financial needs for the ecological transformation are massive. The main two reasons are that countries have to increase their renewable energy production and that many sectors have to be electrified in order to become emission free. Estimates about the necessity for investments are in the range of 1.9 % to 2.2 % of GDP by 2030 and an additional 1 % by 2050 for the European Union (Pisani-Ferry, 2022). Thus, in a first stage until 2030, the reduction of carbon emissions will come by adjusting different sectors. The most significant investments are necessary for commercial buildings and residential housing. These alone require investments around five times as high as the necessary investments for energy and transport combined (Pisani-Ferry & Mahfouz, 2023). This underlines why it is so important to discuss the role of finance for the ecological transition of Europe. The EU has implemented many regulations with the aim of greening the financial sector. What is important is that these regulations are interdependent and also intertwined with climate protection measures of the

EU (Brühl, 2021). This is also the approach of this book not to focus exclusively on finance but also to look at other important policy areas such as international relations and innovation that have repercussions for finance as well. The individual chapters of the section of the book deal with different important regulations of the EU including the EU Taxonomy, the Green Bond Standard, the disclosure regulations, and the role of money and fiscal policy.

Chiara Pappalardo's chapter explores the critical role of sustainable finance in driving the EU's transition to a low-carbon economy by 2050, in particular through the framework of the European Green Deal. Sustainable finance, as defined by the European Commission, involves the integration of environmental, social, and governance (ESG) considerations into financial decision-making, leading to long-term investment in sustainable economic activities. These considerations are crucial for advancing both energy and environmental policies. The chapter emphasises the need for standardised corporate practices to reduce greenwashing and ensure regulatory harmonisation among Member States. Despite significant progress, the author identifies several shortcomings in the current sustainability framework, including the risk of excessive financialisation of sustainability and potential regulatory gaps that could hinder the achievement of carbon neutrality by 2050. Further policy adjustments and innovative approaches are therefore needed to support a truly sustainable economy. Initiatives such as the European Green Deal and the Sustainable Finance Action Plan are key examples of the EU's approach to bridging the climate finance gap. The author concludes that the EU's sustainable finance policies reflect an emerging environmental conscience in financial regulation.

The theoretical and regulatory framework established by Chiara Pappalardo for sustainable finance is essential for understanding the development of the green bond market, which Dejan Glavas explores in the next chapter. The international commitment to tackle climate change, particularly as reflected in the Paris Agreement, has led to significant changes in global finance. Green bonds have become an important tool for mobilising capital for green projects. These instruments offer investors a way to finance projects that contribute to environmental sustainability, balancing financial returns with environmental impact. Dejan Glavas discusses the regulatory challenges and standards surrounding this growing market while highlighting the proactive role of the European Commission in promoting these green bonds. The EU green bond market exemplifies the European Environmental Conscience by facilitating investment in green projects and promoting transparency and credibility in sustainable finance. Indeed, the author highlights the EU's proactive role in fostering a market that balances financial returns with environmental benefits, demonstrating both the regulatory and innovative aspects of the EU's approach.

Tomasz Braun's chapter focuses on the innovative financial instruments that promote transparency and accountability within the EU's green agenda. Tomasz Braun examines how these financial instruments, such as green bonds and sustainability-linked loans, are important for systematically integrating sustainability into financial decision-making processes. In doing so, he explores the challenges of transparency and standards enforcement discussed by Dejan Glavas in the previous chapter. Tomasz Braun highlights the importance of robust regulatory frameworks in supporting the growth and the credibility of green finance markets. These frameworks include strong reporting requirements and third-party verification to ensure compliance with environmental standards. By promoting transparency and accountability, these financial instruments help to maintain the integrity and credibility of the sustainable finance market. These instruments promote market confidence by ensuring that investments meet specific environmental criteria, thereby reducing the risk of greenwashing, as highlighted in the previous chapter by Chiara Pappalardo. The author concludes that the rapid growth and development of these markets demonstrates the EU's commitment to aligning financial activities with environmental objectives, reinforcing its leadership in sustainable finance. This commitment is indicative of a broader European environmental conscience that integrates environmental sustainability into the core of financial practices, fostering a market environment where sustainability and financial performance are mutually reinforcing.

Christoph Weber provides an overview of the role of central banks, and in particular the European Central Bank, in achieving climate neutrality. The chapter examines various initiatives such as bank stress tests, the role of market neutrality between highly polluting and greener companies, and the impact of government and corporate bond purchases. The Green Deal sets ambitious targets for reducing greenhouse gas emissions, promoting renewable energy and improving energy efficiency in various sectors. As an EU institution, the ECB also has a role to play in achieving the EU's objectives. As Christoph Weber shows, this is consistent with the ECB's mandate. The chapter highlights the importance of aligning financial activities with sustainability, a theme also explored by Chiara Pappalardo, Dejan Glavas and Tomasz Braun in their discussions of sustainable finance, green bonds and financial enforcement instruments. Christoph Weber's chapter reflects Europe's environmental conscience by demonstrating the ECB's commitment to integrating environmental sustainability into its monetary policy.

Gaël Callonec, Mathieu Garnero, and Noam Léandri analyse the macroeconomic impact of green finance using the ThreeME model. This neo-Keynesian model is designed to evaluate energy and environmental policies. By introducing a financial block into the model, the authors better represent the role of household savings, private banks, and central banks in financing green activities. The model is calibrated to the French economy. The chapter evaluates policies, such as green quantitative easing by central banks, green lending by private banks, and the channelling of household savings into green investments, and finds that these policies are effective in stimulating investment and supporting the green transition. The authors present four energy transition scenarios (frugal generation, territorial cooperation, green technologies, and restorative bet) to achieve carbon neutrality by 2050, illustrating different paths to decarbonisation. A key finding of this analysis is that none of the four scenarios would lead to recessions. Thus, it is possible to achieve climate neutrality without compromising the economy, in line with the triple bottom line of sustainability. This chapter ties in with Chiara Pappalardo's exploration of the integration of ESG criteria into financial regulation, as both emphasise the importance of sustainable financial practices in supporting the EU's environmental goals. The focus on green bonds and financial instruments in the chapters of Dejan Glavas and Tomasz Braun is further extended by the ThreeME model's assessment of green quantitative easing and other financing policies. The authors underline the need for innovative financial mechanisms to achieve the ambitious goals set out in the European Green Deal, as highlighted by Christoph Weber. By integrating these financial blocks and assessing their macroeconomic impacts, the chapter demonstrates the interconnectedness of financial and environmental policies and strengthens the European environmental conscience.

The EU's financial strategies and policies clearly demonstrate an emerging and robust environmental conscience. The EU has made significant progress in integrating sustainability into its financial rules, instruments, and wider economic policies. This integration reflects a comprehensive and forward-looking approach to achieving long-term environmental goals, including carbon neutrality by 2050.

Innovation as a cornerstone of EU's environmental conscience strategy

Innovation remains a cornerstone of the EU's strategy to address environmental challenges (Jordan & Lenschow, 2008). Recent empirical evidence shows that innovation, as measured by patents, reduces CO_2 emissions in the long run (Mahmood & Tanveer, 2024). To achieve the desired ecological transition, a mix of policies that foster creativity (new approaches) and disruption (replacing existing methods and rules) seems to be the most promising approach (Kivimaa & Kern, 2016). This underlines the importance of innovation for the ecological transition of the economy. The second part of this book provides a comprehensive understanding of how economic policies, corporate responsibility, and governance structures contribute to the European environmental conscience, promoting sustainability across different sectors while addressing contemporary challenges such as digital sobriety. The empirical evidence shows that the best approach to achieving the goals of ecological transformation is a mix of different innovation policies, rather than individual measures that are not embedded in an overall innovation strategy (Veugelers, 2014). This justifies that the chapters in this section of the book illustrate the variety of approaches taken by the EU to integrate sustainability into different aspects of governance and corporate practices.

Daniel Feser's chapter examines how German universities are implementing sustainable practices and policies. More specifically, the author explores the role of knowledge transfer offices at German universities in promoting sustainability-oriented innovation (SOI). The author uses a case study approach, comparing two German universities with contrasting approaches to knowledge transfer. The Applied University of Darmstadt has an explicit focus on SOI, while the University of Göttingen aims to support general regional innovation. The innovative approaches taken by these universities, such as energy-efficient buildings and sustainable resource management, illustrate how economic policy at the institutional level can contribute to broader environmental goals. Daniel Feser highlights the role of educational institutions in fostering environmental conscience among students and staff, and in promoting sustainability through research agendas, curriculum development, and campus operations. Indeed, the involvement of students, faculty, staff, and external partners is important in advancing sustainability goals. Collaborative projects and partnerships with local governments, businesses, and non-profit organisations are key. In addition, student-led initiatives drive the sustainability agenda and foster a sense of ownership among the younger generation. It can therefore be concluded that the activities of knowledge transfer offices can contribute to raising awareness and promoting sustainable practices within the university community. This is in line with other research that highlights the importance of regional ecosystems for eco-innovation (Tamayo-Orbegozo et al., 2017). In relation to the environmental conscience, this chapter demonstrates the innovative nature of solutions and grassroots support for effective solutions, two key dimensions of Hörber and Weber's (2022) concept of European Environmental Conscience.

Continuing the theme of innovation and stakeholder engagement, the next chapter by Naciba Chassagnon-Haned and Amina Hamani explores the evolving role of corporate responsibility in promoting environmental sustainability. The authors examine how companies are increasingly integrating sustainable practices into their operations, driven by regulatory pressures, market demands, and ethical considerations. They discuss how companies are moving from linear to circular economy models, which emphasise the reduction, reuse and recycling of materials. This transition is crucial for minimising waste and optimising resource efficiency. Their chapter examines the importance of sustainable supply chain management in corporate sustainability strategies. Indeed, companies are increasingly monitoring and improving the environmental performance of their suppliers to ensure that sustainability criteria are met throughout the supply chain. The authors also highlight the role of technological innovation in driving corporate sustainability, such as the adoption of renewable energy sources and advanced waste management systems. Their chapter provides a comprehensive analysis of how corporate strategies and governance structures are being adapted to address environmental challenges. This includes the establishment of sustainability committees, the appointment of Chief Sustainability Officers, and the integration of sustainability metrics into executive compensation. Their analysis puts an emphasis on the growing environmental conscience of companies, reflecting a commitment to sustainability that goes beyond compliance with regulations. Their chapter shows how companies are proactively adopting innovative solutions and engaging stakeholders to achieve environmental goals.

Maxence Tétard and Marjorie Tendero explore the concept of digital sobriety and its relevance to sustainable governance. Digital sobriety refers to the conscious effort to reduce the environmental impact of digital technologies by promoting more sustainable practices in their use and management. The rapid growth in the use of digital technology has led to increased energy demand, often met by non-renewable energy sources, thus exacerbating greenhouse gas emissions. Digital sobriety involves several key practices such as optimising digital infrastructure for energy efficiency, reducing unnecessary data storage and transmission, and prolonging the lifespan of digital devices through maintenance and upgrades. The chapter outlines practical steps and policies that can be adopted by individuals, organisations, and governments to achieve digital sobriety. Raising awareness of the environmental impact of digital technologies is crucial. The chapter highlights the importance of education in promoting digital sobriety. It also discusses public campaigns and corporate initiatives aimed at educating users about the benefits of digital sobriety and how to implement it in their daily lives. These include campaigns to reduce digital waste, promote recycling of electronic devices, and encourage the use of energy-saving settings on personal devices. This chapter contributes to the concept of European environmental conscience by emphasising the need for sustainable digital practices. While aligning with Hörber and Weber's (2022) dimensions, this chapter uniquely highlights the integration of digital sobriety into sustainable governance, reflecting an innovative approach to managing the environmental impacts of the digital age. The emphasis on education and awareness promotes grassroots support for effective solutions and demonstrates how the informed and conscious use of technology can significantly reduce environmental footprints.

The EU's environmental conscience in its external relations

The EU has long been recognised as a global leader in environmental governance. The EU seeks to extend its influence and ambitious environmental policies standards beyond its borders. The EU's environmental conscience is therefore manifested through its external relations. Indeed, through international agreements, strategic partnerships, and funding, the EU promotes environmental protection and sustainable development on a global scale. Its leadership is characterised by its ability to reconcile ambitious internal policies with realistic external strategies. In the face of growing environmental challenges, such as climate change and biodiversity loss, the EU is positioning itself as a key player on the international stage, able to mobilise and influence other actors for effective collective action in this field.

Frauke Pipart's chapter investigates the evolution of the EU's ambition in international environmental conventions, with a particular focus on the Convention on Biological Diversity and the Basel, Rotterdam, and Stockholm Conventions over the period 2012–2018. Frauke Pipart assesses the EU's position at the Conferences of the Parties for each convention. The study shows that the EU's ambition has changed very little over the studied period. The EU's ambition is highest in the Rotterdam Convention, where it consistently advocates strict controls on hazardous chemicals and pesticides. The EU pushes for the inclusion of new chemicals in the Convention's annexes and promotes the Prior Informed Consent procedure to ensure that countries are fully aware of the risks before importing hazardous substances. The EU is also ambitious in the Stockholm Convention on Persistent Organic Pollutants. In the Convention on Biological Diversity, the EU shows moderate ambition, promoting the conservation of biodiversity, the sustainable use of biological resources, and the fair sharing of the benefits arising from genetic resources. In the Basel Convention, the EU's ambition is relatively low. This Convention deals with the control of transboundary movements of hazardous wastes and their disposal. The EU's internal policies strongly influence its positions in international negotiations. When strong internal regulations are in place, the EU tends to push for similar or more ambitious international agreements. This alignment ensures that EU Member States do not face conflicting regulatory requirements and can operate within a unified framework. This consistency is an expression of the EU's Environmental Conscience and demonstrates its commitment to high standards both domestically and internationally. Conversely, in the absence of an internal policy framework, the EU's ambition tends to be lower. The political and financial implications of the issues, as well as their legally binding nature, influence the EU's level of ambition. Highly political issues and those with significant financial implications tend to result in lower ambition. In the absence of strong internal policies, the EU tends to adopt more conservative positions in international negotiations. This cautious approach stems from the need to avoid overcommitting to international standards that may be difficult to implement domestically without an established regulatory framework. This reflects a pragmatic aspect of the EU's environmental conscience, balancing ambition with feasibility.

The chapter by Paiman Ahmad, Muhammad Usman, and Atif Jahanger provides a detailed examination of the challenges and opportunities of the European Green Deal for fossil fuel-dependent economies, with a particular focus on Iraq. The European Green Deal aims to achieve climate neutrality by 2050, which requires significant reductions in fossil fuel consumption and increased reliance on renewable energy. This chapter analyses how the transition to a low-carbon economy in Europe is affecting oil-dependent countries and explores the potential for these countries to adapt to the new energy landscape. Indeed, this transition poses serious economic challenges for fossil fuel-dependent economies such as Iraq, where revenues from oil account for a major portion of government revenues. The authors emphasise the need for Iraq to diversify its economy and invest in renewable energy to mitigate these risks. The author suggests, for example, exploring the opportunities of solar energy in Iraq. The author stresses the importance of international cooperation and support from the EU to ensure a just transition for fossil fuel-dependent economies. This fair transition allows the EU to extend its environmental conscience and support global partners in their efforts to combat climate change and develop sustainable energy solutions.

Perspectives on the European environmental conscience in the EU economy

The chapters in this book illustrate how this environmental conscience has penetrated and developed in different aspects of EU governance, from finance and innovation to external relations. These aspects are particularly compelling because these policy areas are not traditionally associated with sustainability debates (Lenschow, 2012). However, they highlight the significant impact that growing environmental concerns are having on various policy areas. In this context, our contribution is pivotal. By drawing attention to the intersection of sustainability and these different policy areas, we not only broaden the scope of environmental discourse, but also pave the way for innovative solutions. Our work underscores the need to integrate sustainability considerations into all facets of policymaking, demonstrating that environmental responsibility is a critical component across the board, not just in traditionally "green" sectors, such as renewable energy, conservation, and waste management. Our holistic approach ensures that environmental considerations are embedded at the heart of policy development, ultimately leading to more sustainable and resilient societal outcomes.

The European Environmental Conscience, as explored in this book, is therefore not a static concept but a dynamic force that continues to shape the EU's political landscape. It reflects the growing awareness and concern among European citizens about environmental issues, and it is a testament to the EU's commitment to addressing these challenges. This environmental conscience has been a driving force behind European integration. The chapters in this book have shown that the integration of environmental concerns into the three policy areas analysed – finance, innovation, and external relations – has been crucial in shaping the EU's approach to sustainable development.

The aim of this was to analyse whether there is a European Environmental Conscience in the EU economy. Looking ahead, it seems clear that a European Environmental Conscience does exist. The answer is yes. This European environmental conscience will continue to evolve and the intersection of finance, innovation and external relations will remain crucial in driving this evolution. However, in the light of recent events, the EU faces significant environmental challenges that require innovative solutions and strengthened governance. For example, the recent energy crisis has exacerbated geopolitical tensions with Russia (Akchurina & Della Sala, 2023, p. 202; Krickovic, 2015). The shift away from dependence on Russian fossil fuels to renewable energy sources has become even more urgent. While the European Environmental Conscience provides a promising foundation for addressing these challenges, its resilience will be tested in the coming years. The European Union's ability to maintain its commitment to sustainability in the face of geopolitical pressures, economic uncertainties, and the urgency of the climate crisis will determine its future.

The finding that a European Environmental Conscience is developing is very important in the light of the results of the recent 2024 European Parliament elections. With additional seats for far-right parties and climate-sceptic politicians, there is a risk that the European Green Deal will be diluted. After a period when sustainability was very high on the agenda, thanks to protests by Fridays for the Future and other initiatives, there has been some backlash against climate policies recently. This makes it all the more important that there is a deep foundation for sustainability in the economy and society. The next few years are crucial to make real progress towards the goal of climate neutrality by 2050. As this book has shown, the focus is shifting from traditional areas such as energy and environmental policy to finance and innovation. This is a good sign, as it will be essential that all sectors of the economy contribute to achieving the goal of climate neutrality.

Bibliography

Akchurina, V., & Della Sala, V. (Eds.). (2023). *The European Union, Russia and the post-Soviet space: Shared neighbourhood, battleground or transit zone on the new Silk Road?* Routledge. https://doi.org/10.4324/9781003081722

Barnes, P. M., & Hörber (Eds.). (2015). *Sustainable development and governance in Europe: The evolution of the discourse on sustainability* (1. issued in paperback). Routledge.

Brühl, V. (2021). Green finance in Europe—Strategy, regulation and instruments. *Intereconomics, 56*(6), 323–330. https://doi.org/10.1007/s10272-021-1011-8

Hörber, T. C. (2013). *The origins of energy and environmental policy in Europe: The beginnings of a European environmental conscience*. Routledge.

Hörber, T. C., & Weber, G. (Eds.). (2022). *The European environmental conscience in EU politics: A developing ideology*. Routledge.

Jordan, A. J., & Lenschow, A. (Eds.). (2008). *Innovation in environmental policy? Integrating the environment for sustainability*. Edward Elgar.

Kivimaa, P., & Kern, F. (2016). Creative destruction or mere niche support? Innovation policy mixes for sustainability transitions. *Research Policy, 45*(1), 205–217.

Krickovic, A. (2015). When interdependence produces conflict: EU–Russia energy relations as a security dilemma. *Contemporary Security Policy, 36*(1), 3–26. https://doi.org/10.1080/13523260.2015.1012350

Lenschow, A. (Ed.). (2012). *Environmental policy integration*. Routledge. https://doi.org/10.4324/9781849771238

Mahmood, H., & Tanveer, M. (2024). Do innovation and renewable energy transition play their role in environmental sustainability in Western Europe? *Humanities and Social Sciences Communications, 11*(1), 1–9.

Matláry, J. H. (1997). *Energy policy in the European Union*. Macmillan Education UK. https://doi.org/10.1007/978-1-349-25735-5

Pisani-Ferry, J. (2022). The missing macroeconomics of climate action. *Greening Europe's Post-Covid-19 Recovery, Blueprint Series, 32*.

Pisani-Ferry, J., & Mahfouz, S. (2023). *The economic implications of climate action*. Unpublished. https://www.piie.com/sites/default/files/2023-11/2023-11-08-pisani.pdf

Tamayo-Orbegozo, U., Vicente-Molina, M.-A., & Villarreal-Larrinaga, O. (2017). Eco-innovation strategic model. A multiple-case study from a highly eco-innovative European region. *Journal of Cleaner Production, 142*, 1347–1367.

Veugelers, R. (2014). *What innovation policies for ecological transition? Powering the green innovation machine*. WWWforEurope Working Paper. https://www.econstor.eu/handle/10419/125726

Index

Pages in *italics* refer to figures, pages in **bold** refer to tables, and pages followed by "n" refer to notes.